Intelligent Buildings

Intelligent buildings should be sustainable, healthy, technologically aware, meet the needs of occupants and business, and should be flexible and adaptable to deal with change. This means the processes of planning, design, construction, commissioning and facilities management, including post-occupancy evaluation are all important. Buildings comprise many systems devised by many people and yet the relationship between buildings and people can only work satisfactorily if there is an integrated team with a holistic vision. This book introduces the concept of intelligent buildings that meet these criteria.

Derek Clements-Croome is Professor Emeritus in architectural engineering and was Director and founder of the MSc Intelligent Buildings Programme in the School of Construction Management and Engineering at the University of Reading, UK. He founded the CIBSE Intelligent Buildings Group in 2006 and in 2009 the *Intelligent Buildings International Journal* which he edits. He is coordinator for the CIB Commission W098 Responsive and Intelligent Buildings. Recently he was awarded a D.Sc honoris causa by the University of Debrecen, Hungary, where there is a distinguished Department of Building Services Engineering.

Intelligent Buildings

An introduction

Edited by
Derek Clements-Croome

University of
Reading

SKANSKA

HILSON
MORAN

Buro Happold

Routledge
Taylor & Francis Group
LONDON AND NEW YORK

earthscan
from Routledge

The University of
Nottingham

UNITED KINGDOM · CHINA · MALAYSIA

First published 2014
by Routledge
2 Park Square, Milton Park, Abingdon, Oxon, OX14 4RN

Simultaneously published in the USA and Canada
by Routledge
52 Vanderbilt Avenue, New York, NY 10017

First issued in paperback 2020

Routledge is an imprint of the Taylor & Francis Group, an informa business

British Library Cataloguing in Publication Data
A catalogue record for this book is available from the British Library

Library of Congress Cataloging-in-Publication Data
Intelligent buildings: an introduction/edited by Derek Clements-Croome.
    pages cm
  Includes bibliographical references and index.
  1. Intelligent buildings. I. Clements-Croome, Derek, editor of compilation.
TH6012.I575 2013
  696–dc23   2013009877

ISBN 13: 978-0-367-57632-5 (pbk)
ISBN 13: 978-0-415-53113-9 (hbk)

Typeset in Sabon by
Swales & Willis Ltd, Exeter, Devon

Contents

Illustrations

Figures

Tables

Contributors

Barış Bağcı, University of the Ryukyus, Okinawa

Derek Clements-Croome, University of Reading*

Gareth Davies, Director at Icon Print Innovations

Martin Davis, Specialist Engineering Alliance; founding director of Integrated Project Initiatives Ltd, the delivery vehicle for the pilot IPI projects

Richard Everett, Intelligent Builders Limited*

Francesca Galeazzi, Arup Associates

Lisa Gingell, t-mac Technologies

Ken Gray, Performance Building Partnership*

Alan Johnstone, Trend Control Systems*

Philip King, Director at international, multi-disciplinary engineering consultancy, Hilson Moran*

Lam Yuen-man Bonnie, Hong Kong College of Technology

Lau Po-chi, Asian Institute of Intelligent Building, Hong Kong

Peter McDermott, Building Integration Consultant Buro Happold*

Ziona Strelitz, University of Reading

Mark Worrall, University of Nottingham*

* Member of the CIBSE Intelligent Buildings Group.

Preface

This book is an introduction to intelligent buildings and was conceived and planned by members of the Chartered Institution of Building Services Engineers (CIBSE) Intelligent Buildings Group which was founded by me in 2006. The book is not an official publication of the CIBSE. In addition we invited some other distinguished professionals to participate. The book has the support of the International Council for Research and Innovation in Building and Construction (CIB).

Each chapter provides a basic knowledge to enable a strategic view of the design and management issues involved to be formed by young professionals and students across all built environment disciplines. The last six chapters present a range of case studies by Buro Happold; Arup Associates; Atelier Ten; ZZA; Trend; and authors from Hong Kong and Okinawa. For those readers wanting more detail they should refer to the *Intelligent Buildings International Journal* and the second edition of the book *Intelligent Buildings: Design, Management and Operation* published by ICE Publishing in 2013.

I would like to thank our sponsor Skanska; Nicki Dennis and Alice Aldous of Taylor and Francis; in addition to the above mentioned companies my fellow CIBSE Intelligent Buildings Group authors: Richard Everett, Gareth Davies, Ken Gray, Alan Johnstone, Philip King, Peter McDermott and Mark Worall. In addition, Martin Davis, Lisa Gingell, Ziona Strelitz, Barış Bağcı Francesa Galeazzi, Lau Po-chi, Lam Yuen-man Bonnie joined us.

Intelligent buildings are a vibrant part of the sustainability debate and they are a major feature in the many eco cities planned in various countries. Above all they affect the working and leisure lives of people. We hope you will enjoy this knowledge taster on a topic that stretches back into history but also forward to the future.

Derek Clements-Croome
Professor Emeritus in Architectural Engineering
School of Construction Management and Engineering
The University of Reading

Abbreviations

ACE	Association for Consultancy and Engineering
ASHRAE	American Society of Heating, Refrigeration and Air-Conditioning Engineers
BEAMA	British Electrotechnical and Allied Manufacturers' Association
BERR	Department for Business, Enterprise and Regulatory Reform
BSRIA	UK Building Services Research and Information Association
CIB	International Council for Research and Innovation in Building and Construction
CIBSE	Chartered Institution of Building Services Engineers
CIC	Construction Industry Council
DCLG	Department for Communities and Local Government
DCMS	Department for Culture, Media and Sport
DEFRA	Department for Environment, Food and Rural Affairs
DTI	Government Department of Trade and Industry
FETA	Federation of Environmental Trade Associations
HVAC	Heating, ventilation and air-conditioning
ICT	Information computer technology
NAC	National Audit Council
OGC	Office of Government Commerce
RIBA	Royal Institute of British Architects
SEC	Group Specialist Engineering Contractors' Group
WLC	Whole life cost

Part I

Principles of design and management practice

Chapter 1

Introduction

Derek Clements-Croome

An intelligent building is one that is responsive to the occupants' needs, satisfies the aims of an organisation and meets the long term aspirations of society. It is sustainable in terms of energy and water consumption and maintains a minimal impact to the environment in terms of emissions and waste. They are also healthy in terms of well-being for the people living and working within them and are functional according to the user needs.

(Clements-Croome, 2009)

Intelligent buildings should be sustainable, healthy, technologically aware, meet the needs of occupants and business, and should be flexible and adaptable to deal with change. The life cycle process of planning, design, construction, commissioning and facilities management, including post-occupancy evaluation, are all vitally important when defining an intelligent building. Buildings comprise many systems devised by many people, yet the relationship between buildings and people can only work satisfactorily if there is an integrated design, construction and operational team possessing a holistic vision.

Buildings affect people in various ways. They can help us to work more effectively, they can present a wide range of stimuli for our senses to react to, and they provide us with the basic human needs of warmth fresh air and security. Intelligent buildings are designed to be aesthetic in sensory terms, including being visually appealing; they are buildings in which occupants experience delight, freshness, a feeling of space. They should integrate daylight into their design, and should provide a social ambience which contributes to a general sense of pleasure and improvement in mood.

If there is to be a common vision, it is essential for architects, engineers and clients to work closely together throughout the planning, design, construction and operational stages of the building's total life cycle. This means that planners, consultants, contractors, manufacturers and clients must share a common vision and set of intrinsic values, and must also develop a single understanding of how patterns of work are best suited to a particular building when served by the most appropriate environmental systems. A host of technologies are emerging that help these processes, but in the end it is how we think about achieving responsive buildings that matters. Intelligent buildings need to cope with social and technological change and are also adaptable to short-term and long-term human needs; however, from the outset this must be delivered through a vision and understanding of the basic function and character of the building.

Throughout history clean air, sunlight, sound and water have been fundamental to the needs of people. Today, sensitive control of these needs may use either traditional or new solutions, or a blend of these, but we have to remember that the built environment is fundamental to mankind's sense of well-being and it is the totality of this idea that we need to understand and value even in this low carbon economy age. Intelligent buildings respect these values for the individual, the business organisation and for society, and we can learn a lot about intelligent buildings by looking at the history of world architecture and seeing how people have adapted buildings to deal with the rigours of climate and the changing face of civilisation. There are also lessons from Nature; animals and plants that have evolved to use materials and expend energy optimally in a changing and dynamic environment. Similarly buildings are now having to absorb the impact of the technological age, but the implications of climate change and the need for healthy working conditions are now also dominating our thinking as people become more knowledgeable about their environment.

Modern buildings consume a great deal of energy and water in their construction and during their total life cycle operation. They use large quantities of materials and aggregates and generate waste and pollution at every stage of their existence. It is no longer acceptable to consider a building and its systems in isolation from its social impacts. The growth of megacities to over 10 million people by 2050 is now part of a rising trend towards urban living and development. Modern liveable cities do comprise intelligent and sustainable buildings and infrastructures, however, they are designed to show a respect for the natural environment in all respects. *Sustainable and Intelligent Cities* are composed of *intelligent buildings* supported by *intelligent infrastructures* and are created for the well-being of the residential, commercial and industrial communities which inhabit them.

In the future, there will need to be a consideration towards the influences buildings have on society, the local community and future generations. For this, we will need to consider the environmental, social and economic impacts of each building throughout the total processes of bringing it into being or deciding to refurbish existing ones. *Whole-life value* in which quality and whole-life costs are assessed is therefore paramount if we are to think long term and meet growing sustainability demands. However, this does not mean architecture has to be starved of human considerations; after all improving the quality of life is an essential ingredient of sustainable development.

For intelligent buildings to be sustainable and to sustain their performance for future generations, they must remain healthy and technologically up to date; they must meet regulatory demands; they must meet the needs of the occupants; and they must be flexible and adaptable enough to deal with change. Buildings inherently contain a variety of systems devised by many people, and yet the relationship between buildings and people can only work satisfactorily if there is *integration* between the supply- and demand-side stakeholders as well as between the occupants, the systems and the building envelope. *Systems thinking* is an approach essential in planning, design and management, together with the ability to create and innovate whilst remaining practical. The ultimate objective should be *simplicity* rather than complexity, which not only requires technical ability, but also requires the powers of interpretation, imagination and even intuition as part of the building process. Building Regulations can stifle creativity yet are necessary to set minimum levels of expectation and to satisfy basic health and safety requirements. However, we should aim our work well above these prescriptive requirements; after all, buildings form our architectural landscape, they generate the environment we inhabit and they should uplift the soul and the spirit of the people within them as well as those who visit them.

The key criteria for achieving good-quality intelligent buildings are defined by Strelitz (2006) as:

- satisfying stakeholder objectives and needs;
- meeting social and environmental needs; and
- recognition of available resources.

An intelligent building starts with a good client brief and should comprise of:

- a clearly articulated project vision;
- a recognition of the planning, design and procurement realities; and
- about whole-life value objective.

The creation of shared visions, effective teams, clear structures and robust processes ensures that the intelligent building being constructed will demonstrate the purpose for which it was conceived. Times are certainly changing so there needs to be an outlook by the project team which is long term and not just short term.

Key issues for intelligent buildings are sustainability (energy, water, waste and pollution); innovation such as the use of information and communication technology, robotics, embedded sensor technology, smart-materials including nanotechnology; adaptability and flexibility; health and well-being in the workplace; and an understanding of social change. Aesthetics as well as function are important in the sense that not just the visual appearance is considered but how the environment affects all the human senses constitutes a total aesthetic.

Some examples of innovation are, for example, coating and embedding materials with nano-particles allowing us to specify material properties much more easily. Such materials in façades, for example, will provide sophisticated forms of feedback and high levels of control besides regulating heat losses and gains. Self-healing materials will revolutionise façades in the future (see *BBC Focus Magazine*, 2009). Another example is Lotusan paints applied to façades to repel water and so let rain water be collected easily. Simple, natural materials can also be effective. In India, Vetiver grass is made from the root of the Vetiver plant and is used as a blind which can be dowsed with water in hot weather to enable adiabatic evaporative cooling to occur.

The intelligent buildings control markets are strong worldwide even after the gloomy economic period of 2009. The largest markets are in the USA, Asia, Middle East and Europe but some smaller countries are showing rapid growth. BSRIA Member e-News August 2009 shows that Scandinavia, Germany and Qatar spend most per capita on sophisticated intelligent controls. The increasing demand for sustainable, healthy and low-carbon intelligent buildings seems, therefore, likely to sustain this dynamic market pull.

Building management systems provide control and interoperability between the various systems servicing the building. Innovations such as internet-based, communication standards and protocols are increasingly making it more important to integrate systems within intelligent buildings. This, in turn, will ultimately require an extended range of professional expertise that could force a cultural change towards the permanent adoption of intelligent buildings.

Terminology

A lot of terms are used. Should we refer to intelligent or smart buildings? Then there is the sentient building (Mahdavi, 2006) which describes how well the building responds to the occupants' changing needs. Mahdavi stresses that a sentient building should be measured continually with a sensor network which can predict and also activate change according to circumstances. Figure 1.1 shows the terms in common use. Automation aspects and high technology, especially in information and communications technology, are the smart elements which are important but a building needs also to respond to social and environmental factors and this is articulated by the language of low-tech passive environmental design. An intelligent building increases the environmental socio-economic value.

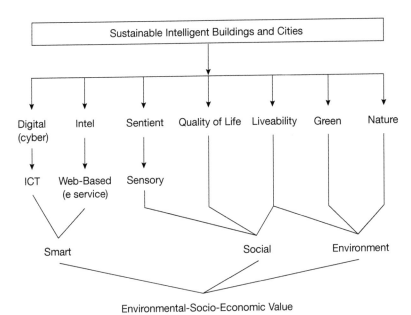

Figure 1.1 Classification of terms

Chapter 2

Building owner's perspective

Mark Worall

Why should a building owner or developer invest in an intelligent building? In general terms, a developer seeks short return on investment (ROI), however, for owner operators whole-life value is the key aspect of an intelligent building. CABA (2008) reported that the cost of an intelligent building should not be significantly higher than traditional buildings and, in the cases examined, showed that the whole-life cycle cost of an intelligent building is lower than traditionally constructed and operated buildings.

In a new building, the initial cost of the project covers the design and build phase of its life cycle, however, minimising resource consumption at the design stage by using passive as well as active methods, as described in Chapter 8, can significantly reduce HVAC requirements and so reduce both the capital and operating costs (HEEPI and SUST, 2008). The minimisation of plant and equipment has the additional benefit of freeing up lettable space or reducing the building footprint, which again reduces cost and improves ROI. Cabling can be significantly reduced because of the integration of power and control systems, such as HVAC, lighting, security, fire and safety.

Johnson (2007) reported savings of approximately $2 per square foot on installing efficient cabling by reducing the cable pathways from between nine and ten to between two and three, saving about $1 per square foot per year thereafter. Combined IT/telecommunications cabling could save 38% on material and labour costs, an integrated building management system (BMS) infrastructure could save up to 33% on material and labour costs. Hirigoyen (2009) showed that rental values could be increased by 2–7%, energy cost savings of between 10 and 50% could be achieved and occupancy rates could be increased by 2–6%.

In existing buildings, refurbishment is driven mainly by energy costs and business change, with payback periods shortened due to upgrades of existing plant and improved control through modern BMS:

- A sports and leisure complex in Toronto with an annual electricity bill of $3M, reduced energy costs by 76% after implementing an automated lighting system. An ROI of less than two and a half years was achieved (CABA, 2008).

• Intelligent control of existing HVAC equipment could reduce running costs by up to 30% (CABA, 2008).

For typical UK building types, the most cost-effective refurbishment options in terms of payback periods, in order of shortest to longest, were (Quartermaine, 2009):

1. variable speed heating pumps
2. energy-efficient lighting (T5)
3. DC fan coil units
4. heat recovery
5. high-efficiency boilers.

Intelligent buildings provide a high-quality environment for tenants and occupants, which contributes to enhanced productivity and reductions in absenteeism and staff turnover. Landlords or developers may not necessarily directly benefit from the well-being of a tenant or occupant, but a high-quality property will result in a more fully occupied building, with a higher lettable value, a reduction in churn, and an enhanced corporate image.

In most studies on refurbishment, it is observed that upgrading the fabric, such as improved insulation, glazing, air tightness and significant changes in HVAC, provides payback periods of over 25 years and so is normally beyond the budget of most building developers. However, ROI can be increased if all the stakeholders that benefit from a refurbishment invest in it. In owner-occupied properties, increases in productivity can result in increases in quality and output:

• Increase in productivity of over 8% occurred in a refurbished post office building when lighting conditions were improved (Innovest, 2002).
• Factory owners can show increases in productivity (CABA, 2008).
• Students' performance can be improved by as much as 20% (CABA, 2008).
• Retailers can gain increase in sales per unit footfall (CABA, 2008).
• Hospitals can discharge patients up to 21/2 days earlier (CABA, 2008).

Legislation and moves to address environmental pollution, energy consumption and carbon emissions are becoming important considerations for owners and developers. In the UK, 2010 Building Regulations (Building Regulations, 2010) require at least a 25% reduction in Target Emission Rates (TER) compared to Building Regulations in 2000 (Building Regulations, 2000a, 2000b). Every building constructed, sold or let is now required to possess an energy performance certificate (EPC) that rates its energy consumption and carbon emissions. The Building Regulations set minimum standards for most aspects of building design and performance but, increasingly, buildings are being designed to meet national and international benchmarks. In the UK, the Building Research Establishment Environmental Assessment Method (BREEAM) (BRE, 2008) is the most well known, where a building is scored across various aspects of its design, construction and operation, including:

- energy consumption;
- carbon dioxide emissions;
- environmental pollution;
- water management;
- environmental responsibility;
- health and safety;
- staff well-being.

By its definition, the designers of an intelligent building consider all of these aspects together, and so owners who commission intelligent buildings are investing in properties that meet current and future needs. However, intelligent buildings are also designed to be adaptable to changes in building use, work patterns, advances in technology, climate change, legislation and public concern. Therefore, an intelligent building will have a longer life span, its lettable space will be more fully utilised, the income generated will be maximised, and thus a greater ROI will be realised.

Chapter 3

Why should a client invest in an intelligent building?

Richard Everett

A client or developer should invest in an *intelligent building* because it is the right thing to do. It is right because intelligent buildings are *economic*, they are *efficient* and they are *effective*; they are more economic, more efficient and more effective than buildings constructed traditionally. Evidence of real buildings in real contexts proves the *efficacy* of the intelligent buildings approach, particularly in the increasingly challenging modern economic circumstances. These advantages are referred to as the 4 Es.

Table 3.1 describes work which shows how intelligent buildings can reduce capital and running costs, energy consumption, and carbon emissions but can also increase productivity, rental value, occupancy and staff retention rates. Thus the building contributes to higher income growth.

How intelligent buildings can improve service delivery and deliver responsible property investment

The need to build new buildings is to improve the quality of the environment for building users, consequently leading to greater success for organisations in reaching their goals. In other words there is a business case for intelligent buildings and one that is rooted in improving the quality of building delivery. It is not about having pretty buildings that win awards, it is not about raising the status of leaders in their peers' eyes, it is not even about making our political masters happy – it is about improving the quality of the environment within which organisations operate so that people are healthier, have a good sense of well-being and work more effectively. About 90% of the ownership costs are the salaries of the staff so it is important to create conditions that let them work as productively as possible.

Intelligent buildings have the potential to offer highly productive and responsible environments that utilise integrated technology to deliver high-quality experiences for building users. However, there are many barriers to this implementation, in particular the inability of some areas of the sector to embrace the holistic and joined-up collaborative approach that is necessary

Table 3.1 Typical examples of the 4 E advantages of intelligent buildings

The four Es	Examples
Economic	Reduction of 24% capital cost (Bowen, 2005) Return on Investment (ROI) of 10 years (Kelly, 2008) Command more rent (Burr, 2008)
Efficient	Reduction of 36% in running costs (Bowen, 2005) Energy bills reduced by 20% (Johnson, 2007) Lighting control reductions of some 30–40% (Ratcliff, 2008) Identity and Access Management strategy paid for out of efficiency gains (Tizard and Mockford, 2008) Energy cost savings between 10% and 50% (Shapiro, 2009)
Effective	Small productivity gain (0.1–2.0%) large effect (Woods, 1989; Clements-Croome, 2000b, 2005) Reducing temperature – higher productivity – 1.8% for every 1°C (Niemelä et al., 2001, 2002; Wargocki et al., 2006) Increase in quality of learning as a result of higher productivity (Everett, 2009; Bako Biro et al., 2007, 2008; Clements-Croome et al., 2008; Wargocki and Wyon, 2007) 0.5% productivity increase pays back within 1.6 years (Wyon, 1996) 17% improvement in productivity – RAE (McDougall et al., 2002) Increased rents by 2–6% (Eichholtz, Kok and Quigley, 2009) Occupancy rates 4.1% higher (Burr, 2008) Higher income growth over 10 years (Baue, 2006)
Efficacious	Sustainable environmental approach (Edwards, 2002) Carbon saving strategies (Carbon Trust, 2002) 15% reduction in global carbon emissions (Thomas, 2009) Focusing on quality of learning (Everett, 2009)

4 Es taken from Carder (1997), Akhlagi (1996) and Himanen (2004)

for success. There is much evidence in many traditional design teams that the various contributors prefer to work in their own silos and seldom work coherently together. But with a well led and highly performing integrated team approach, intelligent buildings can offer a considerable return on investment.

Another barrier concerns financial arrangements and their revision by moving away from a capital-only perspective which only looks at project costs, towards one that looks at the potential lifetime cost (i.e. whole-life costing). Equally, new financial models are required that move away from the reliance of institutions to source money from overstretched capital streams and towards increasingly innovative finance options based upon carbon funding. Add to this an underpinning desire to apply a sustainable approach to lessen the carbon impact of constructing and operating buildings – intelligent buildings can contribute to all of these issues and achieve high productivity and sustainability returns. This return, however, can only be attained if the cultural and process barriers so often seen in traditional design teams are resolved. A whole-life value approach is essential so that optimum whole-life costs and quality are both ensured.

So how do intelligent buildings add value to this agenda? Ask some people about the environmental agenda and they would say that Information Communications Technology (ICT) is part of the problem – in this context, however, it is part of the solution. This concept was acknowledged by the Joint Information Systems Committee (JISC, 2009: 5) in its draft strategy for 2010–2012: 'ICT is not part of the problem; it is not a cost burden. ICT is part of, and a big part of, the answer.'

So we see that 'intelligent' buildings promise reductions in running costs. The former achieved through infrastructure efficiencies, and the latter, through reduction in energy bills. Add to that the more productive workforce and this all adds up to being very sustainable indeed. Further advantages are the higher rental values and lower churn rates that can be achieved. So, even if the intelligent building has a higher capital cost, the values offset this and payback periods are usually less than three years.

'We cannot afford to allow the building practices of the past to be maintained if they result in inefficient, ineffective, uneconomic buildings' (Everett, 2009). We have to prioritise to make sure that we are building economically; we need to think about the efficiency of the building process in capital and revenue terms – together, these amount to considering the full-life costing of the building. Another question also has to be asked: "Is this the right thing to do?" It is here that some of the competing ideals come in and a 'balance' has to be made. However, when selecting that balancing point the business objective (i.e. improving the quality of business delivery) has to be paramount.

Therefore, without the essential ICT infrastructure in place, the automation, control and systems processes cannot be realised. It is therefore essential that ICT consultancy advice is sought in the strategic planning of the building, otherwise the 4E examples in Table 3.1 will not be achieved.

The rationale is also an attractive one to developers and lenders and innovative financial arrangements are achievable, especially if one considers the possibility of outsourcing and gaining economies of scale with collaboration across an area, particularly within consortia or crowd sourcing arrangements. Responsible property investment is increasingly becoming more common with many companies entering the market in 2009–10. This trend is as a result of greener assets being more desirable for investors (Hirigoyen and Newell, 2009).

So, the overall life costs of the building, including its running costs, rather than just associated with its construction phase, need to be brought into account. Traditional financing options do not do this. Typically the ROI or Return on Capital Employed (ROCE) is measured ignoring the running costs or support costs of the building – this therefore distorts the analysis and means that the true savings to be made by utilising an 'intelligent building' approach are hidden.

The benefits identified in Table 3.1 add up to a powerful case for taking into consideration the whole-life cost of buildings, rather than just the design and construction costs often considered by many traditional design teams. For significant inroads to be made into the sustainability agenda whole-life costing must be used. The following figures emphasise the point because typically, during the lifetime of a building with a capital cost of £10 million, it will cost around £50–100 million to operate and, more importantly, £2 billion for business to operate within it. So, if we are to address the ROI and the effect on the environment properly, we need to address the cost and environmental impact not only during construction but over the whole lifetime of the building, using a whole-life value approach.

Hopefully the obvious benefits of a converged and integrated approach to constructing buildings will lead to a new era of innovative intelligent buildings that are cost efficient, socially responsible and sustainable. Not to address this would be 'tantamount to wasting money' (Everett, 2009) or as Victor Fairey (Clements-Croome, 2004) puts it, 'In the world of intelligent buildings, this is financial negligence'.

Conclusions

In these increasingly difficult financial times we need to make the best of our limited resources. This means gaining savings on operation costs as well as the cost of construction of our buildings. It means the best staff productivity. It means designing with the end result in mind – in this case of increasing the quality of service delivery. And, most of all, it means doing the right thing. We need to make sure that we are doing this for all the right reasons – not just to satisfy one agenda but to balance the competing agendas so that they are all satisfied. In short building an intelligent building creates a win-win situation all around.

Chapter 4

Bringing intelligence to buildings

Gareth Davies

The effective delivery of 'Intelligent Buildings' will require the installation and maintenance of integrated information systems (intelligent Building Management Systems – 'iBMS') that can manage a multitude of varied operational processes and collate data from these processes for supervisory examination.

To better understand how this might be achieved then we might:

1 Examine a highly functional 'intelligence system' – in this case the human autonomous nervous system, to identify characteristics that might act as pointers to an optimised logical 'iBMS' model.
2 Consider some lessons from the revolution in commercial Information Technology of the last 30 years that has brought us from the personal computer to the ubiquitous mobile device and see whether some of these ideas can be applied to the delivery of the characteristics we have identified.

A sophisticated control system – our own!

Fortunately for most of us most of the time, the Autonomic Nervous System (ANS), classically divided into two sub-systems the Parasympathetic and Sympathetic, remains a mystery. It can remain invisible and untended, quietly going about its business for many decades perhaps emitting the occasional alarm, adapting itself to our changing circumstances. Occasionally we measure the vital signs, checking its function and the devices for which it assumes responsibility. It seems able to collate signals from multiple sources to deliver a coherent response to complex environmental stimuli.

This apparently simple and yet incredibly sophisticated 'organ' exhibits many of the characteristics that we would seek in an 'iBMS'. That is:

1 We normally do not know it is there.
2 It tells us when something is wrong.

3   It does not need a lot of maintenance.
4   It changes the way it works depending on changing circumstances.
5   It can provide important diagnostic and condition information to drive any remedial action or performance enhancement.
6   It can use 'sets of sensors' to deliver a coherent response through 'multiple actuators'.

This seems a pretty good start given that the system performs its operational functions as required. Can we identify any design or engineering features that might support these characteristics?

The ANS can be morphologically described as a discrete entity connected into a central processing unit – i.e. the brain – and to the other bodily 'organs' with which it interacts. These organs can be thought of as separate functional

Parasympathetic System

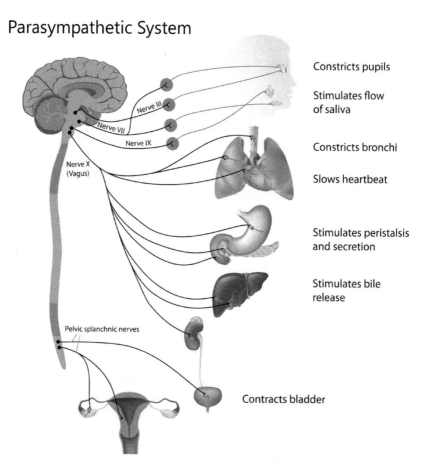

Figure 4.1 Parasympathetic Nervous System

systems, 'black boxes' that act as 'sensors' and 'actuators' or as sets of a combination of the two. They can be re-connected and re-configured in the case of, say, loss or malfunction, e.g. prosthetic limbs or in a heart transplant. A 'brain' gives us the coherent picture and the ability to adapt to changing circumstances, for example, through training and education or reacting to damage or the wear and tear of actuating parts. In the human system, we do not have two brains in case one breaks down but obviously in an iBMS this need not be the case. Once we reach adulthood, it is reasonable to assume that our adaptive behaviour is driven more by changes to 'software', i.e. brain connections, rather than the underlying 'hardware' of the brain, though clearly this is a matter of debate.[1] In a similar manner, it can be said that common protocol is used to send messages through the system – nerves seem to work in the same way in the legs and the arms.

Sympathetic System

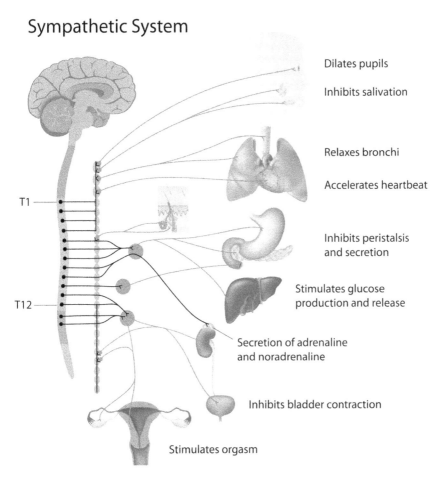

Figure 4.2 Sympathetic Nervous System

Any metaphor can become laboured and stretched but consider some ideas that have emerged:

1   A brain is needed to interpret a variety of complex input signals to return a coherent adaptive response.
2   The various elements of the system can be treated as 'black boxes' with internalised functionality that respects a particular set of inputs to deliver a given set of outcomes.
3   The use of a stable, homogeneous infrastructure upon which a common protocol is carried.
4   That adaptability and flexibility can be independent of the underlying infrastructure hardware.

An Information Technology revolution

In the early 1980s, computing power had begun to migrate out of the air-conditioned room towards the hand, driven by exponential improvement in hardware performance and subsequent price reduction (articulated in the industry by the oddly named 'Moore's Law'[2]).

Companies such as DEC and Data General had grown out from mainframe suppliers and IBM was now selling mini-computers with partners providing the software functionality. In the UK, ICL was following a similar business model.

Many other technologies and practices emerged to best utilise this dramatic advance in hardware performance:

1   Formalised software engineering methods arose in response to the need to reduce the cost of software development and of adapting existing software to changing requirements. These methods introduced 'black boxes' and modularity to systems design.
2   Early 'open source' platforms were built so that algorithms and business rules were encapsulated within software independent of the underlying infrastructure – software and hardware were separated. This gave software the quality of an 'asset' as it could be 'ported' forward through generations of supporting technology rather than re-written for every advance.
3   New ways of storing data were devised, for example relational and network databases, intended to simplify the task of constructing information from data. These databases could be accessed via a common 'language', thus providing a homogeneous data platform for any system comprising a number of varied functional elements.

With the arrival of the IBM PC, and Apple and Microsoft beginning their battle for dominance of the desk-top, these trends aligned to create the tsunami of change responsible for the IT ecosystems of today.

It can be seen that some of these ideas are analogous with some of the features found in our 'human model'; let us consider one or two in more detail and how they might be applied to the delivery of effective intelligence to building management.

Black box, not a black hole

'Structured design' and 'structured programming' were responses to the exploding cost of re-engineering software code written using methods for the programming of early computers. These methods were inadequate for the expanded software repertoire enabled by more powerful computing devices and produced software that could not be maintained over any non-negligible period of time. The greatest challenge faced by designers of information systems then, as now, is that the system as originally envisaged is, inevitably, not the system that will be functioning in even perhaps the shortest of terms.

How then to design a system for change?

The idea is that systems should be built out of 'black boxes' where the initial design focuses on the logical solution which can then be translated into physical form. With 'black boxes' extra functions can be logically added and then physically implemented and physical mechanisms can be repaired or upgraded without detriment to the whole; truly modular design in other words. This technique is familiar from other manufacturing environments, for example motor cars and the phrase 'plug and play' is now common currency. For a number of reasons these practices are still not universally adopted in the development and deployment of information systems – a good question is why this should be.

An excellent illustration of the 'black box' is the famous 'Turing Test' (Figure 4.3); an essential idea in the discussion of artificial intelligence (http://www.turing.org.uk/turing/scrapbook/test.html). To paraphrase – it is too difficult to determine what intelligence is, so instead the focus should be on the manifestation of that intelligence via the responses to interrogation, i.e. the environment. A system does not care how the 'black box' works, just the responses that it delivers when interrogated or given instructions.

A very interesting point was made during the panel discussion at the CIBSE seminar on the practical issue of what constitutes a logical 'black box' when attempting to make sense of an existing set of building controls systems. A facilities manager referred to an installation where there were five existing control systems and he sought an answer as to how to integrate those systems. In the simplest of models, each of these existing systems might be considered a 'black box'. For example, and for obvious reasons, the optimum practical manifestation of the fire control system may be as a completely self-contained 'black box' that is only, as it were, an 'emitter' of information to the other systems. On the other hand, for the management of energy and carbon emissions, individual devices such as chillers, boilers and air-conditioners might

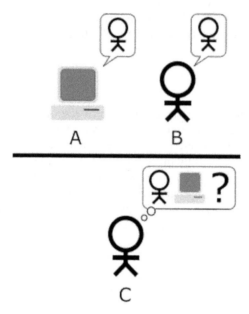

Figure 4.3 An illustration of the Turing Test (http://wikipedia.org/wiki/Turing_test)

be considered as self-contained 'black boxes' that can be wrapped in a new more effective management framework.

In conclusion, the adoption of 'black box' or modular engineering as a design framework is an essential first step for the successful integration of the various control systems required to deliver 'intelligent business' functionality. A major obstacle remains for the controls industry because of the proprietorial ownership of communication protocols. What happens if the 'black boxes' are 'black holes' and cannot talk to each other? 'Garbage In Garbage Out' does not work very well but 'Garbage In, Nothing Out' is no use at all.

Open systems

The adoption of 'open systems' was the IT industry response to proprietorial control of standards. Tim Berners-Lee chose the open TCP-IP protocol for his web work, not the closed proprietary DEC protocol, to support his vision of the open and free exchange of information. There are always exceptions including Apple whose persistence with 'closed' technology was fundamental in their being ousted from dominance in the personal computer market by Microsoft.

With building controls, closed protocols and technology lock-in may have delivered value to a cartel of suppliers but commercial IT history suggests such practices can only occur for a limited period until alternatives arise – 'disruptive' technologies and business processes. If this has not or does not occur then it would be of great interest to try to understand the reasons for the apparently different behaviours of these two markets even as they come to occupy common ground. The use of open protocols between 'black boxes' enables their 'decoupling' so that true modularity can exist (Figure 4.4). Modularity delivers both adaptability and resilience to an information system. By establishing a principle of modularity in iBMS deployment, a framework is established for referral when technical or practical limitations are encountered.

Commodity hardware

It would be hard to deny that the major driver of change in the post-PC world has been the extraordinary increase of hardware computing power available to device designers. This increase has driven down the price of such power, turning the underlying electronics into a commodity. In business computing, the proportions of investment in hardware and software services have shifted significantly towards services; the latest hosting model, 'cloud computing',

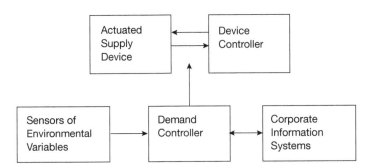

Figure 4.4 An example of a logical black box in an iBMS model

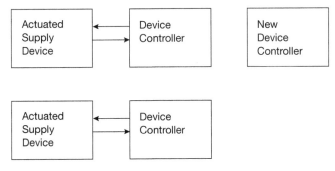

Figure 4.5 How a facility manager might see an iBMS with an example of the enabling technology required

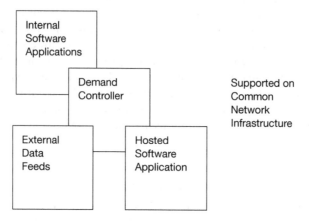

Figure 4.6 A top-level IT view of a set of black-box systems, each of which can be opened like a set of Russian dolls

seeks to reduce the cost to the end-user of hardware and software to a negligible minimum. Even in new consumer sectors, such as smartphones and tablets, where the physical characteristics of the device, its 'look and feel' (more formally the Man-Machine Interface (MMI)) are the main concern, value has quickly migrated to software and services.

In a sense, this progression mirrors our 'human model' where underlying infrastructure is separated from the functions that deliver value to endow the system with maximum flexibility for adaption to change.

Migration of value

Our argument is that the three trends considered have played a crucial role in the explosive growth of information services over the past 30 years or so. Namely:

1 The explosive growth of computing power capacity that has enabled the 'layering' of information system infrastructure.
2 Software Engineering as a discipline focused on the maintenance of software as a productive asset over a protracted period of time, supporting business processes with modular and black-box engineering.
3 The arrival of 'open systems' to facilitate the interoperability of information management systems.

The effect of these trends has been to drive value from hardware to the software application that 'floats' upon a supporting infrastructure and thus drastically reduce the cost of system implementation and maintenance.

This migration has continued remorselessly – who would get excited about a new release of Windows these days? Operating systems and even databases are considered as platforms for the delivery of business value where the battle is to become the de facto standard but the money is in the services associated with these technologies.

The future

In the 1980s Microsoft pushed the pioneering Apple into a siding from which it took nearly 30 years to re-appear. In an echo of this process, Google and others are using the open 'Android' platform to attack the 'closed' Apple business model which remains true to that originally espoused by Steven Jobs. While the brilliance of the design of the iPhone user-interface is undeniable, it has quickly become an established standard that Apple has failed to protect in the courts. In the same way, Microsoft copied Apple's adoption of the GUI and the mouse to erode their competitive advantage. Will history repeat and Apple once again become a supplier to a dedicated and sophisticated niche?

Figure 4.7 Google vs Apple Android vs IOS

Conclusion

It is our contention that the benefits that accrue from 'intelligent buildings' in increased productivity and reduced resource depletion cannot be delivered until controls systems complete the journey undertaken by commercial IT systems over the past 30 years, where value comes to reside first in software and then in associated services. The platform is there in personal computers and the smartphone, commoditised infrastructures for the delivery of information services. What are we waiting for?

Notes

1   Susan Greenfield D.Phil. http://bjp.rcpsych.org/content/181/2/91.full
2   About Moore's Law http://www.intel.com/content/www/us/en/silicon-innovations/moores-law-technology.html

Further reading

Meilir Page-Jones' book entitled *The Practical Guide to Structured Systems Design* (1980, 2nd edn. 1988) is still recognised as a definitive work on systems design. See also http://www.waysys.com

For a fuller discussion of the application of technology to building intelligence, see Michael A. Arbib (2012): 'Brains, machines and buildings: towards a neuromorphic architecture', *Intelligent Buildings International* 4(3): 147–168.

For those interested in Alan Turing, first visit www.turing.org.co.uk, maintained by Andrew Hodges. This website points to Wiki for the Turing Test http://en.wikipedia.org/wiki/Turing_Test.

Recommended is Andrew Hodges' book, *Alan Turing: The Enigma* (2012, London: Vintage) which not only addresses Turing's astonishing thinking but the extraordinary chapter of the Enigma Code.

Chapter 5

Human factors[1]

Derek Clements-Croome

Architecture is more than the art of constructing individual buildings. It is also the creation of environment. Buildings do not exist in isolation. They not only impose their character on their surroundings but also have an incalculable effect on the lives of human beings who inhabit them.

(Conti, 1978)

Introduction

Much research and many surveys have shown that building environments have a direct effect upon an occupant's personal well-being and performance, however, it is only through more recent studies that a clearer understanding of the occupied environment has been discovered. Bako Biro *et al.* (2008, 2012) have shown that primary school children's concentration is affected by the CO_2 levels between 1000 and 5000 ppm and hence the design for effective fresh air ventilation is vital. Satish *et al.* (2011) describes research evidence showing that CO_2 affects decision making even at levels below 2500 ppm. In the UK, the Building Schools Exhibition and Conference (CIBSE, 1999a) asked head teachers if they felt modern buildings affect learning. Around 78% said they felt there was a clear link between the quality of school design and levels of pupil attainment. Williams (2006) reported a similar conclusion for 12 primary schools which he had assessed using the Building Quality Assessment (BQA) method and comparing the BQA score with examination results. *There was a clear correlation between building quality and human performance.*

Miller *et al.* (2009) proved their hypothesis that healthy buildings reduce the number of sick days and increase productivity and also make it easier to recruit and retain staff, by surveying over 500 LEED or Energy Star buildings. More evidence which supports this argument is that showing that sustainable buildings decrease business costs, energy running costs, and as society awareness of green buildings deepens the demand for sustainable buildings increases, so the value of the built asset increases as described in various studies (Thompson and Jonas, 2008; Newell, 2009; Clements-Croome,

2004a). Legislation is forcing the pace. The question is now, can one afford not to be sustainable? Bernstein and Russo (2010) wrote that in the US environmentally labelled buildings rent for 2 to 3% more and have higher occupancy rates and decreased operating costs and, in 2008, building value increased by 10%. Newell (2009) quoted evidence showing that LEED rated buildings cost 6% more to build; have occupancy rates over 4% higher; command 2–6% higher rents; and save 10–50% in energy consumption.

There seems to be a virtuous circle linking health, sustainability and environmental quality. Better building performance is likely to lead to better human performance. Of course, other factors are important such as job satisfaction, social ambience in the workplace and personal issues. Health, here, means mind and body. Our surroundings can influence our moods, our concentration and enhance or detract from our basic motivation to work.

Too often buildings are seen as costly static containers rather than as an investment which if they are healthy and sustainable can add value. Boyden (1971) distinguished between needs for *survival* and those for *well-being*. Human beings have physiological, psychological and social needs. Heerwagen (1998) pinpointed those well-being needs relevant to building design as:

- social milieu;
- freedom for solitary or group working;
- opportunities to develop self-expression;
- an interesting visual scene;
- acceptable acoustic conditions;
- contrast and random changes for the senses to react to;
- opportunities to exercise or switch over from work to other stimulating activities;
- clean fresh air.

In practice investors, developers and clients often agree that sustainable healthy buildings are desirable but want quantified economic evidence to persuade them to finance such projects. Social awareness is changing about the need for sustainable green buildings. The US Green Building Council published a report in 2003 entitled *Making the Business Case for High Performance Green Buildings* and some of the conclusions amended here included:

- Higher capital costs are recoverable in a comparatively short time.
- Integrated design lowers operating costs.
- Better buildings equate to better employee productivity.
- New appropriate technologies may enhance health and well-being.
- Healthier buildings can reduce liability.
- Tenants' costs can be significantly reduced.
- Property value will increase.
- Communities will notice your efforts.

- Using best practices can yield more predictable results but remember occupancy behaviour as well as the quality of the facilities management affects the performance.
- Respect the landscape and open space near the building.

Environmental Factors

In researching impacts of the environment on people it is common to read that environmental factors can act as stressors, so odours, sound, air quality, temperature or light tend to affect humans through four different mechanisms: physiological, affective, stressful, and psychosomatic. Stressors can cause increased heart rate, vomiting, shallow breathing, or muscle tension. They can affect brain rhythms and alter the alpha, beta, and theta patterns, which are correlated with mood and affect. Affective states affect judgement, productivity, interpersonal relations, self-image, morale and aggression. So, one can see the chain of possible physiological and psychological reactions that may occur when exposed to the environment. There are clues here also as to how we may measure the reactions.

We live through our senses and intelligent buildings should be a multi-sensory experience. In general post-occupancy evaluation data show people are very positive about spaces which are airy, fresh, have natural light and views out, preferably to natural landscapes (Clements-Croome, 2006). If an environment is to be conducive to health and well-being it should display the following characteristics:

- a fresh thermal environment;
- ventilation rates to provide fresh air with good distribution and acceptable levels of CO_2;
- good natural lighting;
- no lighting glare;
- appropriate sound levels;
- spatial settings to suit various types of working;
- ergonomic workplaces which have been designed to minimise muscular-skeletal disorders;
- minimum pollution from external sources including noise.

Personal control of these factors wherever possible is important. Central control for items such as security is fine but people prefer to have some degree of control over their immediate physical environment.

Air, warmth or coolth, daylight, sound, space and ergonomics are all important in designing the workplace. However, in the depth of winter or the height of summer temperature tends to be the issue about which workers comment most frequently. There is also another issue because in the current sustainability agenda energy features as a highly important factor and this is closely related to the temperature at which we control our buildings. From a UK survey carried out by Office Angels and the Union of Shop, Distributive and Allied Workers (USDAW) (*The Guardian*, 8 July 2006) the following conclusions were made:

- Heat exhaustion begins at about 25°C.
- 24°C is the maximum air temperature recommended by the World Health Organisation for workers' comfort but note in the UK there is no legislation covering maximum temperatures allowed.
- 16°C is the minimum temperature recommended by the UK Work Place Regulations of 1992 (13°C for strenuous physical work).
- 78% of workers say their working environments harm their creativity and ability to get the job done.
- 15% of workers have arguments over how hot, or how cold the temperature should be.
- 81% of workers find it difficult to concentrate if the office temperature is higher than the norm.
- 62% of workers reckon that when they are too hot they take up to 25% longer than usual to complete a task.

The well-established work on adaptive thermal comfort by Nicol *et al.* (2012) lets the choice of internal temperature be chosen according to the monthly mean temperature. Note the study by Oh (2000) comparing conditions in Malaysian offices with those in the UK showed that people do adapt to temperature but not to air quality. Olfactory reactions to pollutants is similar across countries.

It is worth remembering that people can die in very hot as well as very cold conditions. In more northern latitudes climate change has now brought about as many thermal complaints in the summer as the winter and this trend is likely to continue as seasonal temperatures increase due to global warming.

For fresh air there is the option to have natural ventilation, mechanical ventilation, hybrid systems or air conditioning, but the temperature implications of these systems needs explaining to clients so that they are clear about how many days a year the temperature will be above that recommended. Sometimes clients forget that air conditioning is only effective between certain set design temperatures and outside those it will not work effectively. The USDAW (2006) survey described above assessed that 35% of offices in the UK do not have air conditioning and rely on natural ventilation or fans. Air conditioning, of course, does have disadvantages in that the energy consumption is much higher than for natural ventilation systems, maintenance is much more costly and also there is a slightly higher risk of building sickness syndrome.

Indoor air quality is as important as temperature (Clements-Croome, 2008). Fresh air, like water, is vital to life. A danger with sealing buildings to reduce their energy consumptions is that there will be insufficient air so it is important to build in a controlled air supply such as trickle ventilators or good windows that can be opened a little or a lot depending on the seasonal weather.

Freshness is an under used term in design yet occupants often talk of the need for a fresh environment (Clements-Croome, 2008; Chappells, 2010). Many factors can contribute, such as colour, spatiality and, more often, the air quality. The air quality is a combination of CO_2 level, temperature, relative humidity and air movement. Chrenko (1974) researched thermal freshness votes using a 7-point scale from 'much too stuffy' to 'much too fresh' and found it to be dependent on air velocity and temperature.

Light is reviewed in a report by Veitch and Galasiu (2012); the effects on health are covered in detail. Light has a strong psychological effect on people but it is also linked with views out of the building, colour and spaciousness.

The location of the building with respect to Nature is important. Ulrich (1984) showed how views out from hospital windows on to greenery improved recovery rates. Alvarsson *et al.* (2010) showed that the sounds of Nature aids physiological stress recovery. Greenery and still or running water relieve the body and spirit in any climate.

The surfaces of buildings set the boundaries for *sound*. How a building sounds is just as important as how it looks (Shields, 2003). The shape of interior spaces and the texture of surfaces determine the pattern of sound rays throughout the space. Every building has its characteristic sound of intimacy or monumentality, invitation or rejection, hospitality or hostility. A space is conceived and appreciated through its echo as much as through its visual shape, but the acoustic concept usually remains an unconscious background experience. Libeskind (2002) believes a good building is like frozen music; the walls of buildings are alive. In his words:

> Buildings and systems need to be designed so sound levels are not intrusive to the activities in the space. Façades need to attenuate outside noise from entering the building. However spaces can be too quiet so one has to relate the level to the kind of work being undertaken.

There is increasing electromagnetic pollution caused by mobile phones, computers and other electronic equipment. This is still a rather unknown area with respect to health (Clements-Croome, 2004b, 2000a).Computers can cause eye strain, repetitive strain injuries, posture aches and pains so work patterns need to include 'breaks' so users walk, stand and move around. Desks and chairs need to be adjustable to suit the body build of the person.

Ionisation has always been debated. Nedved (2011) gives an up-to-date account of the knowledge in this area.

The word comfort is perhaps overused. It has a neutral quality. Cabanac (2006) writes about pleasure and joy and their role in human life and indicates how transients are important to give variety for the human sensory system to react to. Well-being is a more comprehensive term. Ong (2013) presents a set of essays entitled *Beyond Environmental Comfort.*

The nature of productivity

For an organisation to be successful and to meet the necessary targets, the performance expressed by the productivity of its employees is of vital importance. In many occupations people work closely with computers within an organisation which is usually housed in a building. Today, technology allows people to work easily while they are travelling, or at home, and this goes some way to improving productivity. There are still, however, many people who have a regular workplace which demarcates the volume of space for private work but is linked to other workplaces as well as to social and public spaces. People produce less when they are tired; have personal worries; suffer stress from dissatisfaction with the job or the organisation. The physical environment can enhance one's work and put people in a better mood, but an unsatisfactory environment can hinder work output (Clements-Croome, 2006).

Fisk (1999; 2000a, b) discusses associations between infectious disease transmission, respiratory illnesses, allergies and asthma, sick building syndrome (SBS) symptoms, thermal environment, lighting and odours. He concludes that in the USA the total annual cost of respiratory infections is about $70bn, for allergies and asthma $15bn, and reckons that a 20–50% reduction in SBS symptoms corresponds to an annual productivity increase of $15–38bn. and for office workers there is a potential annual productivity gain of $20–200bn.

Fisk concludes that there is relatively strong evidence that characteristics of buildings and indoor environments significantly influence the occurrence of respiratory disease, allergy and asthma symptoms, SBS and worker performance.

Measurement of productivity

Often people say this cannot be done but the following four approaches have had success. In their work on the effect of environment on productivity Clements-Croome and Li (2000) propose a holistic model which considers the impact of the social ambience, organisation, well-being of the individual and the physical environment factors and derives relationships between self-assessed productivity and job satisfaction, stress, physical environment, SBS and other factors. This multifunctional approach produced a diagnostic tool which can be used in real-life situations.

Another practical approach is given by Wargocki *et al.* (2006) in which a method for integrating productivity into life-cycle cost analysis of building services is described.

And yet another practical route has been taken on evaluation of productivity by Juniper *et al.* (2009).

Satish *et al.* (2011) used a Strategic Management Simulations (SMS) methodology to measure the impact of environmental factors on thinking, concentration and hence performance.

A reliable methodology is evolving which will produce the evidence we need to convince clients to invest in better buildings which help to improve staff performance and increase value for money, remembering that about 90% of the costs for running a typical commercial office building is the staff salaries.

Sick building syndrome

Many surveys have shown that people can feel unwell while they work in buildings but recover when they leave them. The symptoms usually are rooted in respiratory, eye, cerebral, skin or musculoskeletal discomforts which may exhibit themselves as minor irritations or even pain. Cerebral conditions include headaches and unusual tiredness or lethargy.

SBS is categorized, when 20% of a building's occupants complain of a similar medical condition due to an unknown cause over a period of at least two weeks while in the building (Abdul-Wahab, 2011).

An underlying hypothesis is that SBS is caused by building-related factors. Berglund and Gunnarsson (2000) question this postulate and ask if there is a relationship between the personality of the occupant and SBS. Certainly some people complain about various issues more than others; some people are much more sensitive and therefore much more susceptible to environmental influences than others. They conclude that personality variables can account for about 17% of the SBS variants.

Well-being

The World Health Organisation state 'Health is a state of complete physical, mental and social well-being and not merely the absence of disease or infirmity'.

Well-being reflects feelings about oneself in relation to the world. There is a growing interest in the term well-being with research centres at UK universities of Warwick (Wellbeing in Sustainable Environments (WISE), see Burton *et al.*, 2011), Cranfield (Juniper et al., 2009) and the Institute of Well-Being at Cambridge (Huppert *et al.*, 2005).There is also a notable body of research on well-being conducted at the Health Management Research Centre at Michigan University.

Warr (1998a, b) proposed a view of well-being which comprises three scales: pleasure to displeasure; comfort to anxiety; enthusiasm to depression.

Steemers and Manchanda (2010) propose another definition that encompasses health, comfort and happiness (Chappells 2010).There are job and outside-work attributes which characterize one's state of well-being at any point in time and these can overlap with one another. Well-being is only one aspect of mental health; other factors include personal feelings about one's competence, aspirations and degree of personal control. It is a much more comprehensive concept than the over used word *comfort*. A lack of productivity shows up in many ways, such as absenteeism, arriving late and leaving early, over-long lunch breaks, careless mistakes, overwork, boredom, frustration with the management and the environment.

Work reported in the UK higher education publication *Times Higher Education (THE)* (Newman, 2010) reviewed the impact of well-being on staff and research performance. The Higher Education Funding Council for England is encouraging universities to invest in well-being which can reduce absenteeism and staff turnover. A report by PriceWaterhouseCoopers LLP in 2008 on *Building the Case for Wellness* (commissioned by Health Work Wellbeing Executive, UK) stated that for every £1 spent, well-being brings a return of £4.17. Daly (2010) made a similar evidence case for hospitals.

Well-being is connected with overall satisfaction, happiness and quality of life. It is a more encompassing word than comfort. Well-being depends on the management ethos of the organisation, the social ambience and personal factors, but the physical environment also has a major role to play (Clements-Croome, 2004a). Anderson and French (2010) at the Institute of Well-Being at Cambridge University discussed the deeper significance of well-being. Heschong (1979) reported productivity tends to be increased when occupants are satisfied with their environment. The proposal here is that well-being occurs when all the factors in Maslow's hierarchy of needs are satisfied as shown in Table 5.1.

Table 5.1 Maslow's Hierarchy of Needs in the Workplace (Maslow, 1943)

Need	Achieved by
Physiological	Good working conditions, attractive salary, subsidised housing, free catering.
Safety	Private health care, pension, safe working conditions, job security.
Social	Group relationships, team spirit company sports, office parties, informal activities, open communication.
Esteem	Regular positive feedback, prestige job titles, write-up in company news sheet, promotion and reward.
Self-actualisation	Challenging job, discretion over work activity, promotion opportunities, encouraging creativity, autonomy and responsibility.

Well-being and productivity (Clements-Croome, 2006)

Ten features of jobs are described by Warr (1998a, b) which have been found to be associated with well-being. He believes that stable personality characteristics as well as age and gender are also significant. Environmental determinants of well-being are described as: the opportunity for personal control; the opportunity for using one's skills; externally generated goals; variety; the environment; availability of money; physical security; supportive supervision; the opportunity for interpersonal contact and job status in society. Warr (1998a, b) reviews work that indicates that greater well-being is significantly associated with better job performance, lower absenteeism and reduced probability of the employee leaving. The organisation as well as personal factors are also important.

Heerwagen (1998) draws attention to work in organisational psychology which shows the relationship between buildings and worker performance (P) is interrelated as shown:

$$P = \text{Motivation} \times \text{Ability} \times \text{Opportunity}$$

An individual has to *want* to do the task and then has to be *capable* of doing it; last but not least resources and amenities have to be available so that the task *can* be done. The built environment provides physical and social ambience which affects motivation; the provision of individual control and a healthy environment can enable ability to flourish; communications systems, restaurants and other amenities aid motivation and ability even further by providing opportunity for task implementation.

To understand how we can produce more productive environments means we have to understand more about the *nature of work* and how the human system deals with work. High-quality, and hence productive, work means we need good concentration. When we are about to carry out a particular task we need to settle down, get in the mood and then concentrate. Our attention span usually lasts about 90–120 minutes and then natural fatigue comes into play and our concentration droops, but with a creative break we pick up again, concentrate for another spell of time and this pattern repeats itself during the waking day. This is the so-called *ultradian rhythm*. De Marco and Lister (1987) described this as a concept of *flow*. Mawson (2002) describes their work which claims that individuals take about 15 minutes to ramp up to their concentration level. When an individual is in a state of *flow* they then may be distracted or may become naturally tired and the process repeats itself. Mawson (2002) believes that there is a significant loss of productivity from distraction which has been identified by the *Harvard Business Review*, for a well-managed office, as being approximately 70 minutes of lost productivity in a typical eight-hour day. The distraction is mainly due to general conversation.

There is substantial evidence described by Heerwagen (1998) showing that positive moods are associated with the physical environment and everyday events such as social interactions (Clark and Watson, 1988). Even more telling is the research which showed that positive moods aid complex cognitive strategies (Isen, 1990) whereas negative moods due to distractions, discomforts, health risks or irritants arising from the physical or social environments, restrict attention and hence affect work performance. Because positive moods directly affect the brain processes (Le Doux, 1996), it can be concluded that many aspects of a building's environmental design can aid task performance.

Cao and Wei (2005) described evidence which suggests that low temperatures tend to cause aggression, and high temperatures tend to cause aggression, hysteria, and apathy. The question is then 'Do temperature variations cause investors to alter their investment behaviour?' They hypothesized that lower temperature leads to higher stock returns due to investors' aggressive risk-taking, and higher temperatures can lead to higher or lower stock returns since aggression and apathy have competing effects on risk-taking.

Conclusions

The environment plays a vital part in our personal and working lives. Design needs to recognise this and clients need to be shown that high-quality design is an investment which increases business value. Human factors are fundamental in writing a brief for an intelligent building and it is important that post-occupancy evaluation is carried out to ensure the environmental conditions needed for good work performance and healthy work environments are maintained.

Acknowledgement

I would like to express gratitude for the initial assistance given to me by Gary Middlehurst.

Note

1   A longer version of this chapter appears in the book *Intelligent Buildings* edited by Derek Clements-Croome, published by ICE Publishing in 2013.

Chapter 6

Integration of people, processes and products

Derek Clements-Croome and Lisa Gingell

Integration is a process aimed at achieving coherence, harmony and unity. The opposite is segregation which leads to fragmentation. Integration applies to building design teams; systems which are composed of many products; the process of bringing occupants into the loop as their behaviour influences the performance of systems. A building requires many building services to function, and systems integration allows interaction between them to ensure interoperability. At a strategic level integration requires consideration of the interaction between people, processes and products. The aim must be to achieve integration by an integrated design and management team during the process of planning, designing, constructing and operating building services systems. The underlying direction that dominates the decision to integrate is to achieve substantial reductions in costs and resources but with a more effective operation of the systems for the benefit of the occupants. This will require good monitoring by employing the post-occupancy evaluation techniques which are now emerging. The information gained can be used to improve the quality of new and existing systems.

People, processes and products

Our ability to control temperature, humidity and air cleanliness has made urban development possible in the most inhospitable of locations. Electric lighting, air conditioning and the development of smart glazing systems have loosened the restrictions on building form and fenestration issues. The ability to illuminate the full depth of an office was an important development but this has to be offset with the human need for natural daylight. Because these advances incur embodied and operational energy penalties, buildings need to start with passive design to ensure low energy consumption and limit the capacity needed for active systems; this also lowers the maintenance costs and reduces the plant room space requirements.

Intelligent buildings require intelligent processes. Commercial building processes take place within the confines of the market place but are organised around specific areas of practice that converge, overlap and yet remain distinct. The commercial building industry is a series of linked industries arranged along a *value chain* or *value stream* where each loosely coupled link contributes value to the process. While all the various links taken together form the process, each exists as a somewhat separate social world with its own distinct culture, logic, language, discipline, interests and regulatory demands. By understanding the differences between all the various markets within construction, a new and better understanding of the construction industry itself might be developed (Gray and Flanagan, 1989; Gruneberg, 2000).

The processes in any business organisation are constantly changing, either through legislation, for example the UK Building Regulations 2010 and 2013 Part L, or sustainability demands such as carbon reduction. Companies that have well-documented, easily accessible, end-to-end processes can assess the full impact of these changes on their business model and will be able to adapt them quickly and more consistently. Areas of concern for integration include the need for:

- good planning and briefing at the inception stage of the project;
- coordination of information across the whole building process;
- standardised processes and products rather than allowing the proliferation of proprietary systems;
- interoperability of systems and their interfaces;
- documentary evidence on integrated processes;
- proven and tested processes to be adapted and used on other similar projects;
- mandatory work processes as part of their working patterns.

The way that firms in the construction industry are structured is a result of their own evolutionary history. Beyond specialist niche markets, the vast majority of construction projects remain rooted in a local context. But the construction sector has experienced extensive structural changes over the last three decades, and these changes have had an impact on the way organisations and people in the construction industry carry out their work. The way in which these linked independent practitioners then conceive, design and construct their products will directly affect the way systems are integrated within the services sector, in particular building services systems. We can expect change to continue and this includes not only our state of knowledge but also the nature of our disciplines and the experience of a wider body of stakeholders.

It is essential to have an understanding of the relationship between various practitioners and their differing viewpoints for proper integration to take place. The most important features of a project for the various members of the traditional design and construction team are listed in Table 6.1.

Table 6.1 The views of all the interested parties in the traditional design and construction process need to be appreciated if mutual understanding is to be developed

Practitioners	Viewpoints	Objectives
Users	Social	Usability; natural light, space and fresh air, with some degree of personal control of thermal conditions; amenities; healthy working conditions
Clients	Economic	Reliability, quality; economic operating costs; after care
Designers (e.g. architect, engineer/consultant)	Technical	Overall quality and reliability
Developers/ planners/surveyors	Business/ economic	Conformance to requirements; costs
Contractors	Technical	Quality; profit; workmanship; delivery times
Project managers	Operational	Integration; facilitating resources; coordination; time constraints
Facilities managers	Operational	Operation and maintenance; POE; green issues
Financiers	Economic cost	Successful completion of project on time within budget

Technology and integration

There are three stages in which attention should be focused on technology and integration:

1 design and installation;
2 commissioning and post-occupancy evaluation; and
3 operation and maintenance.

Good integrated buildings work for all stakeholders, something that can be achieved through proper briefing. However one should not forget the technology drivers that can stifle systems integration due to:

• speed of innovation;
• a lack of interoperability across systems; and
• breakdowns which limit the operational availability of the systems.

Intelligent buildings aim for simplicity of operation and this requires the effective choice and use of high technology but with a basic foundation of passive design which uses low technology. It is a blend of high and low technology, which is important to consider in designing an intelligent building.

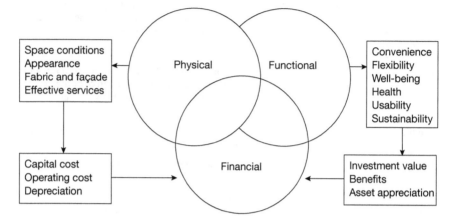

Figure 6.1 The three facets of building performance (adapted from Williams, 2006)

To ensure that integration is achieved, it is essential for a *systems integrator* to be appointed at the start of the project. The systems integrator ensures that all the stakeholders are integrated into the project, and should have the following skills:

- experience of the systems to be integrated;
- the ability to think innovatively;
- logistic skills;
- good communication skills;
- leadership skills.

Integrating human behaviour with building and systems performance

Williams (2006) defines the physical, functional and financial components of performance shown in Figure 6.1 and there has to be a unified view about all these issues.

There is no denying that modern society is fast paced and defined patterns of human behaviour are more and more difficult to distinguish as more individualism is encouraged to open up creativity. Before we can begin to think about integrating human behaviours into buildings and systems designs and management processes we need to look at what drives us as people.

We live in a materialist world articulated by consumerist societies driven by the media and with this come more demands and increased expectations on desired working conditions. Humans are also driven by popular culture, the 'en vogue' of everything we do whether it be work, rest or play. People mostly know their rights and are quick to point this out even if there can be also a loss of responsibility.

We expect our surroundings always to be comfortable and reliable wherever we are. For example, 30 years ago it was perfectly acceptable to have an office temperature set around 15°C. Now employees demand temperatures of 21°C or higher in their work place. This may partly be due to changes in climate, but it is also an indication that expectations have risen above the barometer of 'sit and make do'.

So in a 'demand more' and 'expect more' society, it is more important that we ensure we 'control more', especially when it comes to managing energy consumption, while respecting personal control is desired by occupants and has advantages in terms of productivity.

In former times many controls were not accessible to occupants. Nowadays most people feel competent and confident enough to try to operate them; for example, by overriding controls, but without fully understanding the consequences of their actions and so allowing the air conditioning and heating to run at the same time, or forgetting to change time-clocks for daylight saving or bank holidays. It has to be said that the usability of much equipment is lacking in many cases but there is now research being conducted in this area.

People do not consider how their actions may impact on the overall energy or water consumption of the building. Propping windows and doors open, impacting on the air conditioning or heating levels, or running industrial machinery inefficiently are activities that can lead to further waste but often arise because some aspects of the design or installation are lacking. For example, working in a naturally ventilated cellular office with no thermostatic control or on/off switch means when the occupant is absent energy is wasted and, in the quickly variable UK climate, on warmer days the windows are opened even though the central heating control has not weather compensated or switched off the heating.

It is no longer just the facilities manager who controls a building's system performance, operation and use; it is all the occupants in the building as well. However, clients have to be guided as to how the building works by the design team. Facilities managers should show users the feedback from post-occupancy evaluations and explain the energy and water patterns of use.

So, is integration of human behaviour key to improving buildings and systems performance?

The simple answer to this question is yes, however it is rarely this simple. Therefore, it is important to look at the extent to which human behaviour (i.e. people) can affect buildings and how systems perform.

To demonstrate how people can have both a negative and positive impact on system and building performance it is pertinent to look at an actual case study from 2009 by t-mac Technologies Ltd in Derbyshire, focusing on The City Inn Hotel, Westminster where there were issues with the amount of energy consumed in their conference room – an area which had long periods of both being occupied and unoccupied.

t-mac installed their energy metering system to gain an insight into the metering profile for over 18 areas within the hotel. The sub-metering produced data which are vital to 'showcase' and 'quantify' the building's energy performance and improvement opportunities.

It provides evidence for changing the BMS control strategy and highlights 'misuse' of systems and the inefficiencies or ineffectiveness caused by occupancy behaviour.

For example, the chart in Figure 6.2 details the profile of consumption for the conference room at the outset:

t-mac identified that a high base load profile of energy usage was due to equipment such as air conditioning being left on in the room overnight and the high day profiles were a result of leaving this running constantly overriding local room controls. This demonstrates that people's misuse has a negative impact on energy consumption profiles and ultimately the cost associated with this waste. It is possible that people are unaware of the impact they can have on consumption, hence why feedback sessions are vital to help overcome this. We all need to learn how buildings work.

In this project this is exactly what was done. In order to address the 'people' factor, management arranged for a 'show and tell' session with the staff in which they were shown the energy consumption patterns in the conference

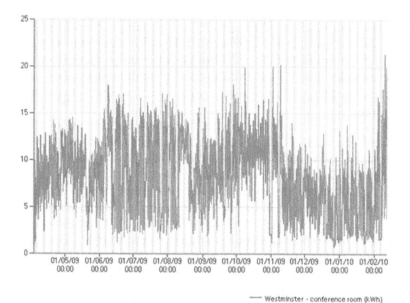

Figure 6.2 Profile of energy consumption in the conference room of the City Inn Hotel, Westminster

room over a period of time and the reasons for it being so high by demonstrating the impact of unneeded equipment being left running overnight.

The session also showed the 'monetary' as well as the 'green' aspect of this wastage and it was collectively agreed that something could and had to be done about this. The awareness of the wastage due to a lack of knowledge was established.

Afterwards an energy management task force committee was organised, with all staff taking an active interest in reducing consumption and cost for the building. It also served as an opportunity to build team spirit by ensuring staff were working together to meet shared targets. On the systems side, the BMS engineer used the metering data, to further 'tweak' the BMS control strategies and staff focused on the primary cause of the wasted energy by ensuring equipment was not left on when not needed.

This process formally began in mid-November, with the conference room a prime focus of the staff's attention, with caution being exercised in the use of equipment, ensuring it was switched off when not in use.

Even after a week the results showed a significant improvement in consumption demonstrating the power 'people' can have on a single electricity metering profile.

Therefore educating, motivating and empowering staff can have a positive impact on consumption and cost. In this case a reduction in the base load of about 11#kWh over 24 hours was achieved (see Figure 6.3).

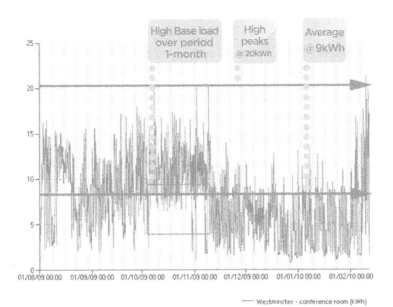

Westminster - conference room (kWh)

Figure 6.3 Educating staff resulted in a reduction in energy consumption in the conference room

Another example of how the 'people factor' can impact on building and systems performance is in relation to Energy Performance Certificates (EPC) and Demonstration Energy Certificates (DEC) ratings. EPC takes into account how a building *should* operate based on the 'assets', whereas DEC takes into account how the building *actually* operates.

A company can carry out an EPC on their building and receive an efficient D grade, but have a DEC on the same building and received a less efficient G grade. While there are other factors involved in the EPC and DEC ratings, the main observation is how 'people' can affect a building's energy rating from the 'desired' to the 'actual'. The impact of occupancy behaviour on consumption of energy or water is a prevalent issue found in many UK buildings including homes. It affects carbon emissions, running costs and systems performance.

Is it possible to influence and change human behaviour?

Most of us are guided by economic drivers, consumerism, popular culture, rising expectations about comfort, but also embedded habits and lifestyles which start shaping us in childhood. Therefore, before embarking on any change within staff an organisation must first look at:

- What do people want?
- Why do they behave a certain way?
- What would appeal to their 'environmental/green conscience'?
- Showing them what changes are possible and the benefits that accrue by implementing them.
- Motivating and praising when a change has occurred.
- Having 'green reward' schemes, of which two examples are now given.

Universities now, for example, have a Green League with awards for the top universities each year. To guarantee success students and staff are all involved in increasing green awareness, for example, by informing all of what equipment to switch off daily, weekly and termly.

In the UK owners of cars which emit less than 100#g/km or less of CO_2 do not pay motor tax and, in addition, those that also meet the Euro 5 standard for air quality can apply for a 100 per cent congestion charge discount as at the time of writing (January 2013).

Another example from t-mac's City Inn case study clearly demonstrates how their Energy Task Force Committee managed to change the duration of the operation procedures conference room.

The month of August 2009 produced the consumption pattern in the conference room shown in Figure 6.4.

Just five months later, in January 2010, the conference room produced the pattern shown in Figure 6.5.

The City Inn project shows how clear improvements were evident after the Energy Task Force Committee interventions; the average base load reduced

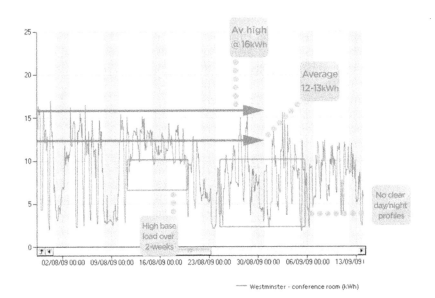

Figure 6.4 Energy consumption in the conference room in August 2009

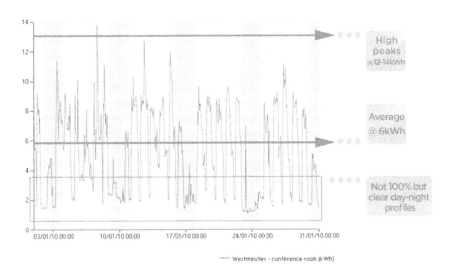

Figure 6.5 Energy consumption in the conference room in January 2010

to around 6 kWh instead of 12–13 kWh and the average daily peaks were 12 kWh instead of 16 kWh.

By educating, showcasing, and empowering people (and adding controls) it is possible to integrate human behaviour with building performance. By encouraging people to be more 'aware' of the environment within which they work and the performance of the systems they use they will take some responsibility for the management of systems too.

Conclusion

The key principles are to start by *understanding* what people want and when, then look to *educate* them on the best way to use systems and *empower* them or let them see the benefits of their actions. Effective facilities management is essential. Manufacturers need to improve the usability of their products.

Chapter 7

Robustness in design

Ken Gray

The design of building services for intelligent buildings should be based on a whole-life performance, which, when viewed with the façade and also occupant feedback provides the whole-life performance of a total facility.

In order to embrace whole-life performance, there needs to be an understanding of the total whole-life picture. The whole-life value of an item or system not only includes the embedded energy in all its components, the manufacturing costs, the processes of design, procurement, installation, replacement and disposal, but also the quality, durability, functionality and environmental impact, including sustainability issues. All the factors that affect each of these considerations influence the whole-life performance of a system to a greater or lesser extent. It is not only a question of how long an item or system will last but also a question of reliability, so one has to predict when will it fail and the impact that the failure has upon the business case.

In order for whole-life performance to succeed, an integrated approach is required to designing, constructing and operating an installation. Processes and procedures should all include adequate inputs from the operational team and not over emphasise the initial capital cost but focus on value for money in the medium and long term. Sustainable design necessitates taking a long-term view anyway. As part of the integrated approach required to control whole-life performance, there is also a need to be aware and understand the interfaces that exist between the building services installations and the façade and, of equal importance, those between the design, construction, replacement and maintenance requirements.

The underlying principle of whole-life performance, reliant on the understanding and approach outlined above, is the management of the risk of premature, or indeterminate, failure. The failure of a component, element or system affects operational availability. In particular, the causal effect of failure has to be determined, such that design and manufacturing processes can be modified to eliminate or significantly reduce the impact from such failures.

The modelling principles required include correct cost identification, reliability analysis, reliability centred maintenance studies, replacement frequency analysis, energy usage and the maintenance requirements. The maintenance requirements may have their roots in traditional methods, however modern advances in manufacturing quality and techniques, along with technologies now being built into products, need to be taken into account.

Money spent on robust, meaning reliable and durable, design can be saved many times over in the construction and maintenance costs. An integrated approach to design, construction, operation and maintenance with input from constructors and their suppliers can improve health and safety, sustainability and design quality. It can also increase buildability, drive out waste, reduce maintenance requirements and subsequently reduce whole-life costs. Off-site prefabricated construction is now much more extensive.

There is now an opportunity for suppliers and manufacturers to supply packaged equipment and plant items with inbuilt 'intelligent' features to provide condition monitoring, maintenance histories, energy usage and availability criteria. The gathering of such data will enhance future whole-life predictions and improve FM response and performance.

Through-life analysis is an expansive concept. The system being analysed is the whole building and its use through time. The services, considered as integrated with the façade, should be designed and optimised at the same time as façade decisions are made.

Architectural features such as building mass, enhanced insulation, solar shading, considered orientation, and the overall envelope shape all contribute to services systems sizing and can be optimised for lowest whole-life cost. This will minimise the sum of the capital cost of the plant, and year-on-year maintenance and energy costs.

Other façade decisions can go further than simply reducing the size of services systems. Beyond a certain threshold of building envelope energy performance the design can enable novel heating and cooling systems, while space assigned for tanks and distribution pipework can enable more self-sufficiency in water supply and sewerage requirements. However, it should also be noted that:

- The process of advancing the design to completion through a few discrete stages does not always produce the best design integration.
- The increasing requirement of whole-life performance analysis for the business case may be intended to help this, but does not necessarily lend itself to option appraisal. Generally there are only two options analysed; the do-nothing and the preferred option.
- Compliance with energy targets could cause an increase in lifecycle cost if more plant is required, but if compliance is achieved largely through façade measures, then plant will be smaller and lifecycle costs will also be reduced. Natural passive cooling and heating methods should be adopted wherever possible.

Flexibility and adaptability

Buildings are generally designed and constructed at considerable financial cost and with considerable environmental impact. It is not only in the owner's interest, but that of society in general, that there is a long-term utilisation of the building. The longer the useful life of the building, the lower the impact of the initial construction cost and the embedded carbon over its lifetime. A notable example of adaptability is the Bankside Power Station in London which began full generation in 1963 but closed in 1981 and then was empty for 13 years before being acquired to become the highly successful Tate Modern art gallery after 1994. Basically the roof of the power station was made into a light box and the internal layout was redesigned.

Flexibility is generally concerned with reconfiguring the internal spaces to accommodate short-, medium- or long-term change. To accommodate these anticipated changes the services design must be similarly flexible.

The planning and distribution of systems must reflect the likely changes in internal layout. These generally comprise of:

- demountable and reusable interior partitioning and suspended ceilings;
- modular furniture/desking that can be used for individual workstations; team work areas, conference tables;
- plant and equipment space;
- additional occupant space.

The services infrastructure must be efficiently sized to accommodate space utilisation requirements and controlled effectively to maintain plant efficiency and availability. Consideration should be given to ring systems for energy and ventilation systems.

Control and lighting systems need to manage energy consumption and the new breed of intelligent systems and objects will greatly assist the integration of 'island' solutions and systems, reducing waste and duplication. The control systems should be able to operate efficiently at both building and workstation level, using small responsive zones to match the activity in the space. Occupation patterns may require small areas in the building to be operational and situations like this must be addressed without the need to 'operate' the whole building.

Controls and building management systems are progressing towards web-based and internet solutions, not only for local control, but for remote control and monitoring of multi-building and disparate assets. The use of such data systems provides a controls infrastructure which is 'plug and play' and will to some extent allow for changes in use.

The advance in wireless technology will also assist in the flexibility and utilisation of space.

Adaptability is concerned with designing buildings for a future change of use. As such it is a long-term consideration. Some key issues in adaptability are:

- Designing the building structure or frame on a grid which can accommodate different building uses and similarly for the floor to floor heights. This is often referred to as 'loose fit design'. The associated building services infrastructure should follow a similar grid layout and distribution where cores and risers may form an integral part of this approach.
- Modular construction, particularly associated with off-site prefabrication. Building elements and often complete 'accommodation pods' can be manufactured off-site with inbuilt services for connection on a 'plug and play' basis. Consideration should be given to services distribution and riser elements which can be manufactured off-site and opportunities examined to integrate building services and structural elements.
- Designing for deconstruction. At some point the building may have to be removed, recycled, refurbished, demolished or decommissioned. This issue is an important part of the whole-life value approach to design and delivery of buildings. Costs and consideration to deconstruction must be included in the whole-life appraisal and business case.

Chapter 8

Passive and active environmental quality control

Mark Worall

Traditionally, buildings have been designed to provide thermal comfort for occupants through close control of heating and cooling systems. BS EN ISO 7730 (1995) defines a standard for thermal comfort based on a heat-exchange model of a human being, taking into account the six factors shown in Table 8.1. Comfort is assessed as a percentage of occupants satisfied or dissatisfied, as no building will satisfy every person. The method rates a space by a predicted mean vote (PMV) or a predicted percentage dissatisfied (PPD). The concentration on thermal comfort has led to higher and higher energy consumption (Pout *et al.*, 2002) but, perversely, to lower occupant satisfaction (London Hazards Centre, 1990). Alternative methods of assessing thermal comfort have been developed based on occupant surveys that take into account how people adapt to their environment in order to achieve comfortable indoor conditions. Adaptive methods are now recognised in standards such as CIBSE Guide A (CIBSE, 2006a) and ASHRAE Standard 55 (ASHRAE, 2010). Occupants respond to the local environment in a number of ways as highlighted in Table 8.2. Buildings that require no heating or cooling, sometimes referred to as free running buildings, have a greater range of comfortable indoor temperatures than controlled buildings that use air conditioning. The comfort range follows the outdoor temperature closely, and a 3°C difference between indoor and outdoor mean temperatures is thought to be the maximum before the indoor temperature becomes uncomfortable. In the UK, it is recommended that overheating should be limited to 1% of annual occupied hours per year above an operative temperature of 28°C.

Thermal comfort is important for the health and well-being of occupants, but deprivation of natural daylight and ventilation, the lack of local control of the environment, the re-circulation of air and build up of volatile organic compounds (VOCs) emitted from modern furniture, fabric and equipment can result in lower productivity, increased sickness and absenteeism and higher staff turnover, according to CABA (2008). Environmental quality encompassing thermal comfort, air quality, fresh air and daylight, views and a connection to the outside world and a combination of automatic and

Table 8.1 Factors contributing to the assessment of
thermal comfort (BS EN ISO 7730, 1995)

- Air temperature
- Radiant temperature
- Air velocity
- Humidity
- Clothing
- Activity

Table 8.2 Adaptive response factors

- Physiological
- Psychological
- Social
- Behavioural

personal control of the environment, can be achieved by a combination of passive and active technologies that are complementary to each other.

Various elements of the building contribute to its environmental quality and its ability to control the environment for the benefit of its occupants. For this section, we define five elements that affect environmental quality and which can play both passive and active roles in its management and control: the fabric of the building; ventilation strategies employed; thermal management, including heating and cooling systems; and control systems.

Fabric

The fabric of a building consists of the roof, walls, fenestration, doors, floors, internal walls, beams, foundations, stairwells and lift shafts, internal finishes and decoration.

Ventilation

Ventilation is necessary to provide fresh air to a space, to dilute and remove particulates, odours and compounds emitted from machinery, furnishings and fabric, to limit the concentration of carbon dioxide, to provide a heat-exchange medium and to prevent condensation. Ventilation can be provided passively, actively or by a combination of both.

Table 8.3 Fabric

Passive	Notes
Multiple glazing	Two or three layers filled with air or inert gas. Many older buildings have single-glazed windows, which can contribute up to half of all heat losses, with U-values (heat transfer per unit area per degree temperature difference) around 5 $W/m^2.K$. Double glazing with air gap reduces U-values to around 2 $W/m^2.K$. Argon filled triple-glazed products are now available with U-values less than 1 $W/m^2.K$.
Spectrally selective glazing	Glass coating that transmits only certain wavelengths of solar radiation, so reducing solar gain. Can reduce daylight transmission and so may lead to higher energy consumption due to increased use of lighting and heat gain from lighting systems.
Internal blinds	Intercepts some of the short-wave solar irradiance and reduces solar glare. However, short-wave radiation is absorbed by the blind and results in solar gain within the space.
Integral blinds	Blinds incorporated between panes of glass so that solar irradiance is absorbed and reflected by the blind before entering the room, reducing solar gain as well as solar glare (Poirazis, 2004). More complex system that requires actuating mechanisms within the glazing unit and so more expensive, potentially less reliable and more difficult to maintain.
External shading Brise-soleil	External features that project out of the building and intercept direct solar irradiance before reaching the windows. Are normally set at a fixed angle, but may be designed to be adjustable. Usually designed at a fixed angle that will minimise solar gain during summer conditions and maximise solar gain during winter conditions. Shading system must be robust in order to withstand local weather conditions.
Thermal insulation	In walls that are largely masonry or concrete, over 30% of heat losses can occur across the walls. Sandwich panelled walls consisting of structural and insulation elements can provide high value thermal insulation in modern buildings. In older buildings, options include cavity wall insulation, internal, and external wall insulation. Advantages include reducing heat losses, ventilation losses and thermal bridging significantly, improving the quality and value of older buildings and reducing the need for new buildings. Disadvantages include relatively high capital cost, reduced internal space when internal insulation is used, change in appearance when external insulation is used, which can lead to conservation issues in 'heritage' and older buildings and disruption to occupants during installation.
Reflective surface	The external surface of roofs and walls may be coated with reflective films to reduce absorption of solar irradiance, and so reduce heat gains in summer and losses in winter.

Table 8.3 Continued

Passive	Notes
Natural roof	Grass or other natural materials (Earth Pledge Foundation, 2005) can be used to reduce heat gains and losses in buildings. Grass and earth provide excellent thermal insulation and so reduce heat losses in winter. Green roofs can provide significant cooling in summer through reflectivity and evaporative cooling through evapotranspiration. A further advantage of green roofs is in stormwater management, through the storage and slow release of rainwater. (Castleton *et al.*, 2010) Disadvantages include the need for structural support for the additional weight, irrigation and drainage systems and specialist maintenance.
Internal thermal mass	Beams, masonry, foundations make up thermal mass. Low thermal mass buildings will experience extreme temperature swings in phase with solar input. High thermal mass buildings will have lower temperature swings and will store and release energy in the fabric, shifting the swings out of phase with the solar input and reducing the maxima.
Phase change materials (PCMs)	PCMs freeze and melt at approximately constant temperatures or in a small temperature range, releasing and absorbing latent heat during the freezing and melting processes. In low thermal mass buildings, PCMs can be used to enhance the fabric internal thermal mass. They can be encapsulated in plasterboard, ceiling panels and other decorative finishes (IEA, 2008). The materials are relatively expensive compared to standard products and so are limited in their application. Should costs decrease, PCM building products may become a feature in many naturally ventilated buildings in future.

Active	Notes
Dynamic façade	Double-skin façades, usually a much larger gap than multiple glazing. Can be either naturally or mechanically ventilated (Poirazis, 2004).
Dynamic insulation	Porous element within envelop that allows air to pass perpendicular to the direction of heat transfer. Still in development stage (IEA, 2008).
Chilled beams/ceilings	Pipework integral to concrete beams and ceilings. Chilled water circulated in beam/ceiling to absorb heat in the space. Surface temperatures should be above the dew point to avoid condensation, and so chillers can be operated at higher temperatures than traditional air-conditioning systems, reducing energy consumption in comparison (IEA, 2008).

Table 8.4 Ventilation

Passive	Notes
Passive ventilation	Buildings can be ventilated naturally, using buoyancy, the prevailing winds or with a combination of both. Natural ventilation depends on many factors as well as the natural driving forces: location, orientation, form, its relationship with the surrounding environment, and its internal layout. (CIBSE, 2005a).
Buoyancy driven	Hydrostatic pressure acts on a body (in this case a building) and the pressure acting decreases with increasing height. If the internal and external temperatures are different then there is a difference in hydrostatic pressure between inside and outside. If openings are introduced, then the difference in pressure will drive ventilation. Internal gains include solar, occupants, equipment and lighting. Glazed atria, stairwells and solar chimneys may be used to induce the stack effect by maximising the difference between internal and external temperatures through solar gain. Natural ventilation is not suitable for handling internal heat gains exceeding 30–40 W/m^2 (CIBSE, 2005a).
Wind driven	When wind flows across a building, it is decelerated and increases in pressure at the windward side. As the wind flows around the building low-pressure regions are created. If openings are introduced across high- and low-pressure regions then this will drive ventilation. Cross ventilation occurs when openings are introduced at opposite sides of the building. Cross ventilation may be limited to depths of less than five times the floor to ceiling height (CIBSE, 2005a), and so deep plate, open plan layouts may be difficult to ventilate naturally.
	Windcatcher/Windcowl type natural ventilation systems intercept wind from towers or scoops mounted on roofs (Su *et al.*, 2011). These may be useful for providing natural ventilation in deep plate buildings, and difficult to ventilate spaces.

Active	Notes
	Active ventilation required for bathrooms, kitchens, food processing areas, industrial processes, garages, high occupancy spaces, spaces with high heat loads, such as mainframe computer and telecommunication suites, close environmental control areas such as clinical spaces and clean rooms. Ventilation can be fully active or mixed mode.
Mixed mode	Mixed mode ventilation generally falls into three main categories, contingency, complementary and zoned (CIBSE, 2005a). Contingency systems are designed to be flexible to changes in building use, and have a strategy for adaptation by either the addition or the omission of mechanical equipment. Complementary systems have integrated natural and mechanical ventilation strategies that operate either concurrently or alternately. For instance, high occupancy areas such as classrooms may require additional ventilation for only part of the

Table 8.4 Continued

Active	Notes
	day. Deep plan offices may not be suitable for natural ventilation but a complementary mechanical system may enable the space to be adequately ventilated with fresh air from operable windows. Zoned systems are designed so that some parts of a building are naturally ventilated whilst other parts are mechanically ventilated, for instance, extract systems can be used in bathrooms and kitchens.
Constant air volume (CAV)	Constant air volume (CAV) systems distribute air at a fixed volume around ductwork that supplies the spaces with conditioned air. Centralised heating and cooling systems are used to condition the air to the required quality and the flows are adjusted to suit the cooling or heating loads.
Variable air volume (VAV)	Variable air volume (VAV) systems maintain a constant air supply temperature, but the flow of air is varied. Modern systems will have intelligent control through centralised building management systems (BMS). Various parameters are monitored, such as temperature, humidity, duct static pressure and carbon dioxide concentration, and the flow of hot water, chilled water, refrigerant, and the distribution of air are controlled to achieve a controlled environment within pre-determined set-points. The majority of hot water and space heating is provided by water or steam boilers fuelled by gas and oil or coal.
Heat recovery	Energy used in heating or cooling a building may be dumped to ambient unless heat exchangers are used to recover some of the heat. It is easier to recover heat from mechanical ventilation systems because heat exchangers can be integrated with the ductwork. In natural ventilation systems, the driving pressures are relatively low, and so the pressure losses associated with air-to-air heat exchangers may compromise the effectiveness of the ventilation system. Windcatcher/windcowls may include low-pressure loss air-to-air heat exchangers.

Thermal management (heating)

Space heating for both domestic and non-domestic buildings requires the greatest energy consumption. Figure 8.1 shows annual energy consumption in commercial buildings in the UK (DTI, 2012). The highest energy consumption remains for space heating at over 50%, followed by lighting at 18% and cooling and ventilation at 12%. According to RIBA (2008), the total space heating load is generally caused by three factors: fabric losses, ventilation losses and thermal bridging. Figure 8.2 shows typical heat losses in UK buildings. The majority of the heat is transferred through the fabric. Leakage of cold air into a building and warm air out of a building can contribute up to a third of the heating load. Thermal bridging is caused by

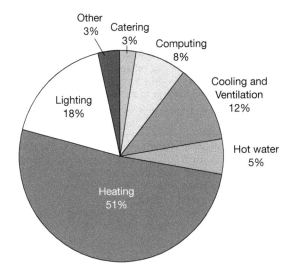

Figure 8.1 UK commercial building energy consumption 2011 (DTI, 2012)

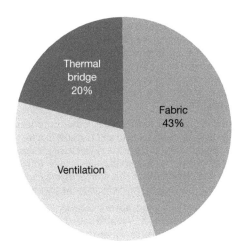

Figure 8.2 Typical heat losses in UK buildings (RIBA, 2008)

building elements that provide paths of high thermal conductivity compared with other elements. Improved insulation of the building envelope, improvements in air tightness and elimination or minimisation of thermal bridging can reduce the heating load considerably.

Table 8.5 Heating

Passive	Notes
Occupants, equipment, solar, lighting	Heat is generated inside the building from occupants, appliances and equipment. Increasing use of IT and related equipment, rising occupant density, and higher air tightness and insulation standards can lead to overheating of buildings in summer.

Active	Notes
Boiler/furnace	Mainly fossil fuelled, seasonal efficiencies of condensing and non-condensing boilers can reach 90% and 80%, respectively. Biomass boilers are increasingly being used. Control of boiler: • modulating flame: burner varied in response to load • step fired: burner changed in steps • modular: multiple boilers brought on line with increasing load • oxygen trim: amount of excess air varied in combustion chamber.
Panel emitters, heater coils	Pumps distribute hot water to feed room panel emitters or heater coils in air handling units (AHUs) and fan-coil units. Modern pumps can run at varying speed, and therefore flow rate can be modified in response to load, increasing efficiency of system.
Combined heat and power (CHP)	Electricity generator driven by engine with waste heat used for space or water heating (CIBSE, 1999b). CHP plants are usually fossil fuelled, with conversion efficiencies of 60–70% compared to 30–40% for engines alone. Small-scale systems (50 kWe–1 MWe) use reciprocating engines, whereas large-scale systems (>1 MWe) use combined-cycle gas turbines or steam generators. Micro-CHP systems (<50 kWe) are coming onto the market for domestic and small commercial applications, using mainly external combustion engines (Stirling engines).
Heat pumps	Heat pumps absorb heat at a low temperature, either outdoors (air source) or below ground (ground source) and reject it in the building at a higher temperature and pressure. Electricity is used to drive a compressor to pump the heat, so may contribute to high energy consumption and carbon emissions. Compressors can be driven at variable speed, maximising their efficiency at part load.
Solar thermal	Flat plate or evacuated tube collectors absorb solar irradiance and provide heating, mainly for hot water. Variable speed pumps can maximise heat output as solar radiation varies throughout the day.
Earth coupling	The temperature 2–3 m below ground remains stable at about 8–12°C and so air drawn through ductwork buried below ground absorbs heat and delivers it to the space. Can be used to ventilate a double-skin façade. Ventilation rates can be varied to suit the demand with variable speed fans.
Fuel cells	A chemical process that combines hydrogen and oxygen to produce electricity, water and heat. The heat can be recovered and used for buildings. A supply of hydrogen is required, either from reformation of hydrocarbon fuels, electrolysis or a dedicated hydrogen distribution system. Potentially over 90% efficiency of converting hydrogen into electricity and useful heat. Fuel cells for buildings are still in the development stage.

Thermal management (cooling)

Overheating of buildings is becoming an increasing problem as buildings improve air tightness, thermal insulation is enhanced, occupation density increases and the use of electrical equipment such as ICT increases.

Table 8.6 Cooling

Passive	Notes
Passive ventilation	Buoyancy or wind-driven ventilation can remove heat gains of up to about 30 to 40 W/m² (CIBSE, 2005a).
Night cooling	At night, when ambient temperatures are low, actuators open to allow air to flow through the building naturally and enable its fabric to absorb some of the heat from the air. As the heat loads to the building increase during the day, the heat stored in the fabric is released, producing a cooling effect. The building requires exposed thermal mass to enable it to absorb and release the heat slowly. It may be ineffective if there are only small swings in temperature between day and night. Mechanical ventilation could be used as a back-up for natural ventilation to assist with night cooling.

Active	Notes
	Fan assisted passive In places where the cooling load exceeds 40 W/m² mechanical ventilation can increase air flow locally and enhance heat transfer.
Mixed mode cooling	Mixed mode cooling involves using mechanical cooling to assist a natural or passive system. As in mixed mode ventilation, contingency, complementary and zoned systems are possible.
Fan-coil	Fan-coil units usually contain heating and cooling coils. The cooling coils are usually fed by centralised chillers.
Chilled beam/ ceiling	See building fabric table. Can be controlled by variable speed pumps and compressors.
Reversed heat pumps	A heat pump absorbs heat from outside and rejects heat inside. The same system can be reversed so that heat is absorbed indoors and rejected outdoors. Variable speed compressors enable systems to run at optimum performance.
CAV/VAV	CAV/VAV systems incorporate cooling coils, normally fed by centralised chillers. Chillers can be closely controlled by varying pump and compressor speeds.
Earth coupling	The temperature 2–3 m below ground remains stable at about 8–12°C and so air drawn through ductwork buried below ground rejects heat and delivers the cooled air to the space. Can be used to ventilate a double-skin façade. Ventilation rates can be varied to suit the demand with variable speed fans.
Combined cooling, heating and power (CCHP)	Sometimes referred to as tri-generation. CCHP systems provide electricity from an engine and use the waste heat from the engine for heating in winter and cooling in summer. Commercial CCHP systems mainly use absorption chillers for cooling in summer. CCHP systems thus utilise waste heat throughout the year.

Control

In the previous tables in this section, the emphasis has been on using passive methods before active methods to minimise the use of energy for maintaining environmental quality, then selecting appropriate technologies to provide ventilation, heating and cooling where necessary. A mix of automatic and manual control of the environment can both minimise energy consumption and enhance occupant comfort. The performance of a building can be enhanced further by intelligent control strategies that use existing and emerging technologies, as described in Table 8.7.

Table 8.7 Intelligent control

Intelligent control	Notes
BMS	A BMS integrates the control of all of the services that a building requires.
Load shifting/ shaving	Coolth can be stored as ice at off-peak periods and used to reduce cooling loads during peak periods. Advantages: prices are normally lower and cooling plant is operated closer to full load.
Demand response control	Demand response control (DRC) (CABA, 2007) was developed as a method of managing energy consumption during rolling blackouts in California in the 1980s. It works by having control systems that reduce demand during periods when electricity prices are at their highest, so reducing cost and demand at peak periods.
Smart metering	Meters can be installed to monitor instantaneous energy consumption and allow a control system to vary lighting, HVAC, etc., in response to loads, changes in weather, spot price of energy. Smart meters will enable facilities managers to identify areas that require attention.
Smart grids	A smart grid is a two-way interaction between energy suppliers and customers using IT systems. It enables peaks and troughs to be managed by energy companies by intelligently managing supply and demand. Increases in renewable energy installations such as building integrated solar photovoltaic systems, windfarms, etc., will require intelligent control between centralised and distributed energy systems. Intelligent buildings may generate their own energy and so may become a supplier as well as a consumer of energy.

Chapter 9

Intelligent buildings management systems

Derek Clements-Croome and Alan Johnstone

Introduction

Over the past 30 years many different buildings have been labelled as 'intelligent'. However, the application of intelligence in buildings has yet to deliver its true potential. Industry has many established intelligent building solutions but it finds it difficult to demonstrate and prove the benefits. Intelligent sustainable buildings improve business value because they take into account environmental and social needs, and occupant well-being, which leads to improvement in work productivity. The ideal system links building, systems within it and the occupants so they have some degree of personal control. Intelligent controls help match demand patterns (Qiao *et al.*, 2006; Noy *et al.*, 2007).

To take an analogy with the brain, you need good quality neural pathways and efficient junctions of these at the synapses but the overall performance is governed more by the connectivity effectiveness between the pathways than anything else. Similarly the technical performance of building management systems (BMS) needs a high degree of integration of different vertical systems and interaction with users. In an integrated BMS, separate systems for each functional area have to be brought together to provide a whole set of solutions of the building control. Often, BMS could not meet the user expectations due to a number of challenging factors:

- The systems may be wrongly specified because of the complexity of multiple stakeholders and possibly conflicting requirements.
- It is difficult to research a consensus of the criteria of optimum performance of the BMS and the buildings.
- The lack of compatibility and interoperability between different systems for sharing information and achieving an overall optimal performance.
- Socio-economic and organisational issues can play a dominant role and can further complicate the design and implementation of the BMS.
- Interdisciplinary collaboration is essential for the nature and complexity of the BMS.

Human orientated management systems

Control systems and communication networks are rapidly developing and in the near future one can expect to see the occupant having more control and interaction with the building and systems. The work of Liu *et al.* (2008), Booy *et al.* (2008) and Qiao *et al.* (2007) shows how by using semiotic modelling this may be achieved by developing a Multi-Agent System for Building Control (MASBO) using semiotic modelling. The primary objective of MASBO is to support the effective management of sustainable energy performance, while taking into account occupant well-being and productivity. This work is discussed in more detail by Gulliver and Liu (2013).

Semiotics, the long-established discipline of signs and information used in social, culture and business settings, has demonstrated great relevance to the built environments (Liu *et al.*, 2008, 2009). From a semiotic point of view, a space for working and living is defined by two important facets: a physical and information space in which its occupants enjoy not only the physical facilities, but enjoy the signs and information. There will be a constant interplay between a building and its occupants, as one is always affecting the other. Once entering a building, an occupant is submerged in a pervasive space in which interactions take place (Duangsuwan and Liu, 2008).

MASBO is designed following semiotic principles which enhance the interaction between the building and its occupants. The epistemic-deontic-axiologic (EDA) architecture, which is developed from semiotics, serves as a theoretical foundation for knowledge representation in software agents.

The epistemic component in the agent keeps the existing knowledge and beliefs the agent attained from previous user requirements and sensed data of the environment. The axiological component represents the basis for value judgements. This basis is formed by the social and cultural conventions, business knowledge and operational policies in an organisation.

The deontic component enables the agent to determine the actions to be performed in response to the external environment. Deontic knowledge, represented as norms in an 'if-then', specifies the relationship between a certain context and expected actions to be taken. A deontic reasoning process can be fired upon receiving a stimulus (e.g. a change of room temperature or a command by an occupant), which may lead to a decision or an instruction for an action (e.g. adjusting the temperature or lighting). The reasoning process will involve all three components by referring to epistemic beliefs, axiological values and deontic norms. The EDA architecture permits the capture of organisational policies on energy consumption and building management, and of individual user's preferences on the control and management of spaces.

MASBO is based on a hierarchical system of agents: personal, local and central agents, as well as agents responsible for monitoring and control of equipment and devices (such as sensors and orthogonal facility systems). MASBO can be dynamically configured with building facilities to meet the requirements for building energy efficiency and personalised work environment, as shown Figure 9.1.

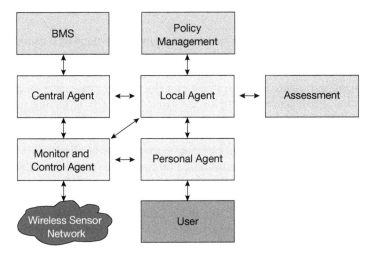

Figure 9.1 MASBO functional architecture

- *Personal agent*: manages user (occupant) profile, observes the work environment, records user's behaviour, forwards operation requests, learns occupant preferences and presents feedback from other agents to the occupant.
- *Local agent*: plays a central role in MASBO. It acts as a mediator, policy enforcer and information provider. It reconciles contending preferences from different users, enforces policies that constrain the environment parameters, provides structural information for their respective zones and responds to environmental state change.
- *Central agent*: has two major functions: decision aggregation and interface to internal/external services required by other agents. The typical services provided by central agent include agent system configuration and interface to BMS.
- *Monitor & Control Agent*: enforces the operation request given by the user, reads and processes sensor data, and creates an environmental state according to decisions made by the local agent.

The multi-agent system acts as an add-on to an existing BMS and requires inputs from policy management and assessment systems. The central agent is the interface for the multi-agent system to BMS. While there is only one local agent for each zone, every environmental parameter for a zone requires a dedicated monitor & control agent. Personal agents are equipped for every occupant in a building working on a mobile device.

MASBO represents the initial outcome of semiotics applied in intelligent buildings and by working closely with experts in the built environment will continue to develop such systems using the latest sensor technology.

Integrated building management systems and their benefits

Much like human beings, intelligent buildings require sophisticated sensory input and multiple systems working in harmony to take appropriate actions or make informed decisions. It follows that intelligent building solutions require an all-encompassing infrastructure and the logical evaluation of a wide array of data. Intelligence also implies the capacity to learn, so intelligent building solutions must deliver post-occupancy continuous improvement through extensive monitoring and effective MI.

This can be delivered by integrated building management systems (iBMS) which are technically advanced solutions based on controllers which are fully configurable, distributed devices, specifically designed to use open protocols on TCP/IP networks with web capability for remote interrogation by web browsers. As technology advances the iBMS can benefit from a more sophisticated array of sensing technologies and a greater degree of integration through convergent open standards.

Integration can occur at different levels: management, control or field. The traditional level model of the CEN TC247 standard, shown in Figure 9.2, defined integration at management level, device level and field level.

This model has now become too simplistic for modern building auto-mation. An enterprise level should be added to cover multi-site cloud-based data storage and applications not really envisaged by TC247 in 1992.

It is also important to differentiate between horizontal integration and vertical integration.

- *Horizontal*: peer-to-peer data exchange in between control or I/O devices – commonly required at the device and field levels.
- *Vertical*: data transfer up from control and I/O devices into a common user interface.

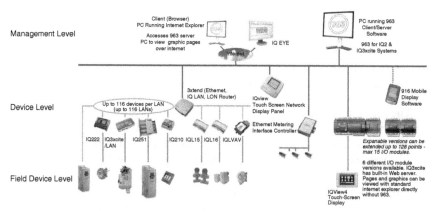

Figure 9.2 Model of control levels

The original intention of TC247 was to define a standard protocol for each level. However, the most common communications standards for buildings controls, BACnet, LONtalk and KNX, all include transmission media and services suited to multiple system levels. LONtalk and KNX were originally designed for lower system levels and primarily support horizontal integration needed for the interconnection of control devices. In addition these protocols both offer a common engineering scheme and KNX has a common engineering tool. In contrast, BACnet provides the best support for vertical integration and, although it supports horizontal integration, the engineering of this is not fully standardised. The OPC standard, developed for the process control industry, also commonly used for iBMS is only applicable to vertical/management level integration.

Buildings can be made more intelligent by better sensing. Installing more sensors can be a costly proposition so manufacturers are developing loop devices and wireless connectivity. Proprietary, point-to-point networks are quite common but several standards are in use. The Zigbee open standard based upon IEE 802.15.4 has been defined for wireless mesh networking in buildings and can be used for interconnection of both controllers and sensors. In addition Z-Wave and EnOcean wireless protocols are used by multiple manufacturers, and gaining market share, for sensors and room control applications. The EnOcean standard is optimised for energy harvesting devices that do not require batteries. The sensors themselves will also become more intelligent, acting as small room controllers, displays or multi-parameter monitoring devices. In the future it is envisaged that sensors may all become internet connected via the 6LoWPAN wireless protocol which supports IPV6 addressing.

Building owners and operators need a better understanding of the utility consumption in their buildings driven by spiralling energy costs, the rationalisation of property portfolios and the demand for carbon neutrality from green investors. Intelligent buildings benefit from better sub-metering using more intelligent metering devices. Most meters are now multi-parameter devices on Modbus or MBus networks integrated with the iBMS. Comprehensive sub-metering presents a much clearer picture of where energy is consumed in a building and aggregated data can be integrated with smart grid and demand response systems.

Complex integration solutions are becoming easier to achieve as a result of the convergence of open standards and IP technology. Typical systems suitable for integration include the BMS, lighting, door access, CCTV, fire and security. The intelligent building will know who is in the building, through door access; and where they are, through RF tagging or CCTV video analytics, and tailor the environment accordingly. Integration is also extending to the fabric itself with solar, PV, natural ventilation, thermal decks, ambient lighting and phase change materials (PCM) likely to become commonplace. We will also begin to see further integration with wider business systems such as Maximo, SAP and Oracle as intelligent buildings strive for optimum efficiency.

While the integration of different systems has become easier to achieve, it is important to consider the human factor. Delivering complex systems requires a fully engaged and integrated supply chain and systems should not necessarily be integrated just because they can be.

The intelligent building designer needs to consider how operators will interact with their building. In other words what data from which service can be beneficially used by other services and what benefits accrue from collating and processing data from different services.

A properly implemented iBMS can allow building systems to perform seamlessly and thus free-up businesses to focus on operational issues. This can have a significant impact on operational cost, which is often underestimated in design. The Royal Academy of Engineering estimates the cost ratio for a new building as 1:5:200 for design and construction, maintenance, and business operation, respectively (Evans *et al.*, 1998). Salaries are typically 90% of the business operation costs. Any system that can help to reduce the

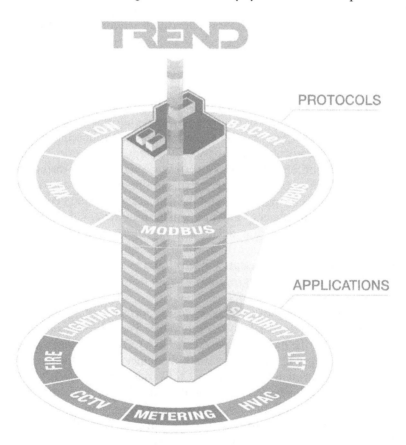

Figure 9.3 The integrated building

business operation elements of the cost base is, therefore, extremely valuable. This can be achieved by increasing staff productivity by providing good environmental conditions in which occupants have increased well-being.

One of the limiting factors for the success of intelligent building systems until now is that they have struggled to engage building users. As users, our expectations of technology have been transformed by the smartphone and the internet. As a result, much of the focus of BMS development is on the user experience with manufacturers providing ever more sophisticated touch screen interfaces, user friendly dashboard displays and web-served content. The BMS is moving away from the plantroom and into the boardroom and reception of our buildings.

Smartphone and tablet technology is developing rapidly and this can only be a good thing for the BMS. Systems often rely on bespoke user interfaces, and while these will improve with the available technology, web-clients and smartphone apps will become more commonplace and have already begun to displace some of the BMS functionality. Using mobile technology to access the BMS helps to maintain comfort conditions at their optimum level by making building data clearly visible for all to see. It ensures that maintenance response is proactive, reducing downtime, and it allows the field operative to communicate effectively with the building user.

Engagement with the user is the most important feature as it highlights how the user and building affect each other and improves the perception of comfort. This can also improve productivity by up to 15% by increasing thermal satisfaction, which is a mix of how comfortable the building occupant feels and how effectively the building operator communicates and responds to issues.

Similarly, the use of web data formats such as XML (eXtensible Mark-up Language) will extend the building intelligence even further, providing a standard format that can be interpreted by IT/data analysis applications. This will connect users more effectively with their buildings and provide lasting benefits.

The data that the BMS collects is extremely valuable and the effective use of the software applications above is only a part of this. Coherent and consistent display of management information allows quick and effective decision making with minimal time to interpret results.

The application of dashboard style displays allows the key parameters for a building or system to be summarised on a single consistent page. The example in Figure 9.4 illustrates how this might be used for a typical office application. Detailed analysis tools are hidden behind the various metrics to allow more in-depth investigation.

Proven control strategies ensure that plant runs effectively to provide a more consistent built environment and alarms are distributed to provide early warning of actionable events to a central bureau or by e-mail or SMS to field operatives, ensuring a proactive maintenance response. Events are stored in

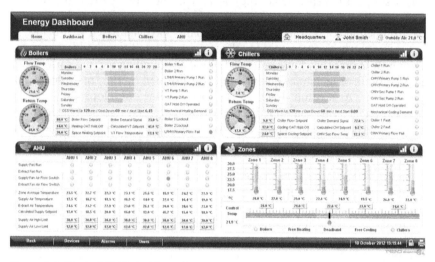

Figure 9.4 Typical BMS dashboard

the SQL database allowing sophisticated data-mining for system and plant maintenance profiling. Systems can be monitored remotely by smart monitoring systems designed to report by exception, meaning that issues are only flagged up when a potential problem is identified, using simple traffic light displays to highlight issues, exception reports for targeted maintenance, and league tables of the poorest performing plant.

The use of these intelligent tools to highlight issues automatically frees up users to concentrate on fixing problems rather than trawling through data. The output is web-served to selected users in a simple graphical format. Consumption outside the expected profile produces an exception which is sent by e-mail to the building operator for immediate rectification. This also ensures that post-occupancy building performance is maintained at optimum levels achieved at handover and provides a platform for continuous commissioning.

Enterprise management of intelligent buildings will continue to grow, driven by Cloud-hosted data and 'software as a service' propositions. This enables the software solutions above but also permits data from distributed systems to be collated and managed centrally. Intelligent choices based on weather or aggregated building consumption can be applied across an entire property portfolio and corporate decisions can be informed with powerful MI about building effectiveness.

Chapter 10

Integrated teams can deliver more for less[1]

Martin Davis

Authority for the principle of integration

It should be axiomatic that buildings that are 'intelligent', coordinated and sustainable can only be created by intelligent people in integrated collaborative teams. The message of integration should have hit home by now. It has been an underlying theme of these authoritative reports:

* Latham, *Constructing the Team* (1994)
* Egan, *Rethinking Construction* (1998)
* NAO report, *Modernising Construction* (2001)
* Strategic Forum, *Accelerating Change* (2002)
* OGC, *Achieving Excellence in Construction* (2003)
* Strategic Forum, *Integration Toolkit* (2003)
* CIC, *Selecting the Team* (2005)
* OGC, *Common Minimum Standards* (2005)
* NAO report, *Improving Public Services through Better Construction* (2005)
* Strategic Forum, *Construction Commitments* (2006)
* OGC, *Construction Procurement Guide to Best Fair Payment practices* (2007)
* Strategic Forum, *Profiting from Integration* (2007)
* Business Vantage/Construction Clients' Group report, *Equal Partners – Customer and Supplier Alignment in Private Sector Construction* (2009)

Now there are drivers: low carbon and an economic crisis

In September 2009, Lord Mandelson announced the Labour Government's *Low Carbon Review*, which had these objectives relevant to intelligent buildings:

* to identify the barriers to improved performance by the UK construction industry and to make recommendations

- to consider how the UK construction industry can take forward the low carbon agenda and to make recommendations to make the UK a world leader in low carbon construction and the built environment,

and a month later a report by Andrew Wolstenholme and Constructing Excellence, entitled *Never Waste a Good Crisis*, urged the supply side of the industry to adopt a vision of sustainability:

> So what will make the industry change now when it has failed to do so before? We believe that an essential step is for suppliers, clients and government to adopt a new vision for the industry based on the concept of the built environment. This means understanding how value is created over the whole life cycle of an asset, rather than simply looking at the building cost, which is only a part of the total equation. It is about how the relatively small up-front costs of design and construction can have such huge consequences for future users, whether expressed as business or social outcomes, as well as for the environment.
>
> [...]
>
> How will this be achieved? We believe that the era of client-led change is over, at least for the moment, and that it is now time for the supply side to demonstrate how it can create additional economic social and environmental value through innovation, collaboration and integrated working – in short, the principles outlined in Rethinking Construction. Clients should focus instead on professionalising their procurement practices to reward suppliers who deliver value-based solutions (Executive Summary)

In reality, suppliers can not change the industry on their own. The time has come for a stronger vision from Government and across the industry which recognises the key contribution that the built environment makes to the UK's long-term economic prosperity and its aim of achieving a more sustainable, low carbon economy (page 15). What we also need are some quick wins that will engage leaders of the top firms, their clients and suppliers, and government. We offer the following:

Industry Leaders

1. Take the lead for the industry's change agenda. Do not wait for clients to give you permission to change. It may be another five or ten years before they will be in a position to help

2. Exploit the recession to look for your case for change – lift the industry by searching for better profits, funded through real value improvements, change and productivity.
3. Seek incentives for delivering innovative solutions. Your customers want them (and need them) but are not sure how to ask! (page 26)

Initiative by the Specialist Engineering Alliance

In advance of these announcements and exhortations the Specialist Engineering Alliance (SEA) comprising, CIBSE, BEAMA, ACE, BSRIA, SEC Group and FETA had been collaborating with 'movers and shakers' from the other sectors of the construction industry to promote the application in practice of the principles of integration and collaboration. The first step was in the form of a report, launched in March 2009, entitled *Sustainable Buildings need integrated teams*:

Sustainability – which at its simplest means improving social and economic outcomes while reducing environmental impact – is the most pressing issue we face in the built environment.

This wide-ranging report . . . draws on evidence from a number of recent publications from BERR, DCLG, DCMS, DEFRA, the OGC and the NAO, more than a dozen case studies, and the views of a number of leading industry figures. The evidence overwhelmingly shows that the procurement process is the major determinant for the sustainability of a facility or construction project.

Delivery processes that are fragmented, hierarchical and adversarial stand in the way of sustainability. Instead more integrated and collaborative approaches are required in which specialists with detailed knowledge of the installation, operation and performance of essential components and systems are brought in at the early stages as part of an integrated delivery process.

Subsequently – with the support of more 'movers and shakers' in both the construction and insurance industries – a proposition was developed by the Specialist Engineering Contractors' Group, on behalf of the SEA, to demonstrate how – through 'live' pilot projects – the following can be delivered:

• improved performance;
• faster programmes;
• greater efficiency;
• less risk.

In parallel, a *Business Case for Integrated Collaborative Working* was under preparation by the Strategic Forum for Construction's Integration Task Group, which demonstrated, with the support of 14 case studies and data from Constructing Excellence, that 'the more integrated and collaborative a team is, the more successful its projects will be and the more benefits they will deliver for all'.

Fundamental to the proposition were the following:

- integrated teams, including the specialists and key suppliers, seconded from their employers, and collaborative working;
- selection by value, not lowest price tendering – in a phrase, 'integrated procurement';
- modern commercial arrangements, including a no blame/no claim culture and gain-share/pain-share;
- independent risk assurance (both technical and financial);
- 'Integrated Project Insurance' (IPI), including for overspend of the agreed cost plan (described in outline later).

The new integrated procurement process

Under integrated procurement the members of the 'core' integrated project team (IPT) of consultant designers, construction manager/contractor, specialist contractors (with any key suppliers with design input), cost advisers and facilities manager, are appointed at the outset. The process of tendering is replaced by the following:

1   Identification of the Business Needs for the project; establish and prioritise the 'success criteria', e.g. low carbon/sustainability, construction time, and investment target. A small advisory team will be formed around the client for the above purpose and the next stage; this will include the end user, an integration facilitator, the risk assurers and insurance brokers. A 'functional brief' will be created, but this will not pre-judge solutions.
2   Using 'selection criteria' based on the above, assembly of the Core Integrated Team (CIC). Selection on the lines of the CIC's 'Selecting the Team' already referenced will be facilitated. This provides an auditable process of scoring (for short-list and then at interview) to select in competition members of the team who should give 'best value' against the chosen and weighted 'selection criteria' and 'functional brief'. Such a balanced competition should yield the 'most economically advantageous' proposals in compliance with Article 53 (1)(a) of the EU Directive 2004/18/EC (and Articles 30(1)(a) and 30(2) of the UK Government's Public Contracts Regulations 2006).

3   Development of design solutions and cost plans by the Core Integrated Team on a fee/time basis; the team will be expanded as necessary so as to include all necessary expertise; and the independent risk assurers (technical and financial) will have open access to this development. All members of the IPT will be focused on finding the best 'value for money' solution, lest they collectively fail the test at 'Gateway 3' below (the stage under OGC's Achieving Excellence in Construction when the investment is approved to go ahead or otherwise); they will avoid using or writing specifications or other 'contractual/protective' documents and letters which have no place in an integrated team culture.

4   Appraisal and benchmarking for investment decision:
- Independent technical and financial appraisal of the Integrated Project Team's proposed design solution and cost plan.
- Comparison with benchmarks of performance for the relevant 'success criteria', making adjustment for any special considerations as appropriate.

These appraisals will be the culmination of the inter-active consultation so far by the independent technical and financial risk assurers, and the resulting joint report will go to the insurance brokers and insurers' panel for the 'Integrated Project Insurance', as well as the client.

5   Agreement of contract: if the Gateway 3 investment decision is favourable, then the key matters for agreement will include:
- The processes of management and collaboration, including the assimilation of other members of the supply chain.
- Gain-share formulae (e.g. low carbon performance, time, cost outcome) and allocation percentages (between the client and the members of the IPT).
- IPI policy terms, premium and excess; the excess will be matched by the pain-share – the allocation of which will be the same percentages as for gain-share.
- The processes by which the technical and financial risk assurers will monitor performance through to completion – when they will be asked to give a 'clean bill of health' for latent defects insurance.
- Establishment of the Project Bank Account, and the arrangements and authorities for its operation.
- Agreement of a form of 'alliance contract' between the client and the IPT.

6   Implementation, with the entire IPT focused on successful delivery: greater efficiency, elimination of wasted resources, time and money, improved performance, and 'soft landings' at completion and proving. In consequence the IPT should concentrate on managing the time and cost of design/construction and, in consequence:
- share in the gain-share yielded by delivery of the 'success criteria', in the pre-agreed allocations; or

- suffer pain-share in the pre-agreed allocations, with the protection of the IPI policy beyond the excess.

Independent facilitation, risk assurance and project insurance

The innovative arrangements for risk assurance and cost overrun insurance cover already mentioned were conceived by the supply side of the construction industry in conjunction with leading insurance brokers in order to overcome the blame and liability culture that inhibits an integrated team from collaborating as a 'virtual company'. A new insurance product, called 'Integrated Project Insurance' (IPI), has been under development to insure integrated teams, in lieu of the traditional Contractor's All Risks, Third Party and Professional Indemnity policies. Team members agree to waive their rights to claim against each other (except in case of fraud) and insurers waive rights of subrogation against them.

Independent facilitation and technical/financial risk assurance are provided, as an integral part of the insurance cost, to promote collaborative behaviours and check the team's proposed design solution and cost plan prior to Gateway 3 investment approval; thereafter the project's performance and risk of cost overrun are independently monitored and insured. A panel of leading insurers is ready to give this cover. The insurance cost (inclusive of the above facilitation, risk assurance and latent defects cover for 12 years) is not expected to exceed the total of the many fragmented premiums traditionally allowed from the top to the bottom of the project team and supply chains.

Under this arrangement the potential losses of both the client and the other members of the IPT are limited to their share of the pain-share. This should facilitate openness about estimates when agreeing the cost plan and also collaboration and flexibility as problems or opportunities for innovation arise. The risk of insolvency is also absorbed, which should facilitate the involvement of SMEs.

Proving the benefits

Adoption of these processes, in place of traditional tendering and the 'liability culture' of traditional contracts, is a prerequisite for the R&D pilot projects to test the proposition. In summary, the savings in time and money that can result should be in cutting these processes and cultures:

- counter-productive estimating and procurement;
- silo protectionism;
- the redundant system of prices, variations and claims.

In his report 'Constructing the Team' Sir Michael Latham, with the support of Reading University, predicted that 30% could be saved in terms of improved efficiency. These savings have yet to impact in any significant terms on the supply side, and hence it is estimated that there is some 20% still to be saved.

In the context of mechanical and electrical services the significance of a move from the 'sequential' to the 'integrated' design process can be illustrated by the comparison in the table below:

Sequential	Integrated
Consultants 'design':	Consultants and specialists together:
• Concept drawings • Specifications • Scheme drawings • Coordinated drawings	• Analyse performance requirements and 'success criteria' (e.g. whole-life cost) • Short-list possible system solutions
Specialist contractors do 'detailed':	• Select equipment to meet performance requirements
• Coordinated drawings • Working drawings • Equipment submittals • Builder's work drawings • As fitted drawings and manuals	• Decide routes etc. compatible with building structure and architecture • Cost options and select solution that best meets 'success criteria' • Share drawing work as appropriate
NB: Coordinated drawings may be the responsibility of either the consultant or the specialist.	• Progressively sign off health and safety compliance

With genuine supply chain management the benefits of integrated collaborative teams should be to:

- get projects under way at least six months faster than under traditional procurement, and then save more time in execution;
- deliver superior performance – e.g. low carbon/sustainability;
- give improved efficiency – in design, on site and 'soft landings';
- facilitate the management and control of risk;
- open the door to dramatic cutting of waste ...

The need for the supply side to deliver 'more for less' is at the head of political agendas, and in the context of the requirement for low carbon construction the *Emerging Findings* of the Low Carbon Construction Innovation and Growth Team have supportive observations at section 3.1.2 and, in particular, in Propositions 3–5.

(3) That the industry, working through a collaborative forum such as Constructing Excellence or the Strategic Forum for Construction, should produce a tighter definition of precisely how an integrated supply chain should come together, what the gains would be, and how the client's position could be protected against cost increases resulting from a lack of competitive tension.

(4) That, as an extension of this, a number of integrated teams should develop a delivery proposal for a suitable building type (such as one or more of the thirty six eco-schools which have been announced), with a view to showing how, given the right procurement and contractual arrangements, a zero or close to zero carbon building could be constructed for the same price as a building built only to current Building Regulations.

(5) That the Chief Construction Adviser and the OGC should work with a public sector department or delivery agency responsible for a rolling building programme to seek to agree a procurement and contractual arrangement within which the above proposition can be tested.

A rally-call to move forward

Only the client can ordain the procurement route for his project. If he decides to appoint an integrated team using the 'modern procurement' approaches already described, then that team (the 'supply side') will be empowered to take full responsibility to deliver to the agreed brief. As was demonstrated by the exceptional projects out of the 14 case studies, the supply side is capable, when empowered, to give strong leadership, and to deliver outcomes in terms of quality (including low carbon), time and cost which exceed expectations.

The supply side, which represents well over 80% of project cost, is also capable of extracting the waste that Latham assessed at 30%, but only if it is liberated from the traditional processes that go with procurement by lowest cost competitive tender. These savings will be reflected in the cost plan that will be insured under the pilot projects proposed, and the team will be motivated to deliver further improvements during execution. The projects will therefore offer clients and any funders far more security than is available under traditional contracting.

It is for the clients to say the word in these terms:

- The projects must be designed and constructed by IPTs using collaborative principles.
- Modern procurement and risk management approaches must be utilised so as to establish a sound basis for going forward.
- At Gateway 3 investment stage, approval will only be given if the above

have been applied and the design solution and cost plan proposed are insured.

Epilogue

Later in 2010 – after the presentation of the paper on which this chapter is based was given – the Construction Minister encouraged the Specialist Engineering Contractors' Group to submit a *fully worked-up proposition* on behalf of the SEA to the Coalition Government. The formalised Proposition set out the integrated procurement process in detail and indicated that an expected 15–20% of total cost could be saved by cutting process waste.

This proposition was one of two 'new procurement models' that were endorsed in the Low Carbon Construction *Final Report* as suitable to 'be taken forward for consideration as the Efficiency and Reform Group of the Cabinet Office develops a strategy for procurement with public sector clients, under the aegis of the Construction Clients Board'.

The *Government Construction Strategy*, published by the Cabinet Office in May 2011, set out an objective to 'reduce costs by up to 20% by the end of this parliament' and, relevantly, announced that

- the *new procurement models* would be further developed for trial *on live projects with a view to rolling out successful practices across government.*
- Integrated Project Insurance would be explored *to support new procurement models.*

In a Cabinet Office *Newsletter* in February 2012 it has been confirmed that the SEA proposition, known as *the IPI procurement model*, will be trailed on pilot projects starting in the first half of 2012. Some pilot projects have been identified against the new procurement models, and more are being sought.

Note

1 This chapter is based on a paper presented by Martin Davis at the CIBSE IB Group seminar 'Systems Integration', 23 February 2010.

Further reading

Business Vantage/Construction Clients' Group (2009) *Equal Partners – Customer and Supplier Alignment in Private Sector Construction* [Online] www.business vantage.co.uk/equalpartners/Equal Partners 2009.pdf.

Cabinet Office (2011) *Government Construction Strategy* paragraphs 2.28 and 2.44 and Action Plan items 6 and 11 [Online] http://www.cabinetoffice.gov.uk/sites/default/files/resources/Government-Construction-Strategy.pdf.

Cabinet Office (2012) *Newsletter* – see *Interim Report* and *Pilot Projects* [Online] https://update.cabinetoffice.gov.uk/sites/default/files/resources/construction%20newsletter%20Feb%202012.3.pdf.

Construction Industry Council/Strategic Forum for Construction (2005) *Selecting the Team* [Online] www.cic.org.uk/services/SelectingtheTeam.pdf.

Egan (1998) *Rethinking Construction* [Online] http://www.constructingexcellence.org.uk/pdf/rethinking%20construction/rethinking_construction_report.pdf.

Latham (1994) *Constructing the Team* [Online] http://nec-discussion.web.officelive.com/Documents/constructing%20the%20team.pdf.

National Audit Office (2001) *Modernising Construction* [Online] http://www.nao.org.uk/publications/0001/modernising_construction.aspx.

National Audit Office (2005) *Improving Public Services Through Better Construction* [Online] www.nao.org.uk/publications/0405/improving_public_services.aspx?alreadysearchfor=yes.

OGC (2003) *Achieving Excellence in Construction* guidance [Online] http://webarchive.nationalarchives.gov.uk/20110822131357/http://www.ogc.gov.uk/ppm_documents_construction.asp.

OGC (2005) *Common Minimum Standards* [Online] http://webarchive.nationalarchives.gov.uk/20110601212617/http://www.ogc.gov.uk/documents/Common_Minimum_Standards_PDF.pdf – since updated: http://www.cabinetoffice.gov.uk/sites/default/files/resources/Government_Construction_Common_Minimum_Standards.pdf.

OGC (2007) *Guide to Best Fair Payment Practices* [Online] http://www.b-es.org/b-es-connections/commercial-and-legal/procurement/ogc-guide-to-best-fair-payment-practices/.

Low Carbon Construction IGT (2010) *Emerging Findings* [Online] www.bis.gov.uk/assets/biscore/business-sectors/docs/10–671-construction-igt-emerging-findings.pdf.

Low Carbon Construction IGT (2010) *Final Report* [Online] http://www.bis.gov.uk/assets/biscore/business-sectors/docs/l/10–1266-low-carbon-construction-igt-final-report.pdf.

Public Contracts Regulations (2006) No. 5 Part 1 [Online] http://www.legislation.gov.uk/uksi/2006/5/pdfs/uksi_20060005_en.pdf.

Specialist Engineering Alliance (2009) *Sustainable Buildings Need Integrated Teams* [Online] http://www.secgroup.org.uk/sea.html.

Strategic Forum for Construction (2002) *Accelerating Change* [Online] http://www.strategicforum.org.uk/pdf/report_sept02.pdf.

Strategic Forum for Construction *(2003) Integration Toolkit* [Online] www.strategicforum.org.uk; click on spanner.

Strategic Forum for Construction (2006) *Construction Commitments* [Online] www.cic.org.uk/strategicforum/pdf/commitments.pdf.

Strategic Forum for Construction (2007) *Profiting from Integration* [Online] www.strategicforum.org.uk under *Reports*.

Strategic Forum for Construction (2010) *Business Case for Integrated Collaborative Working* [Online] www.strategicforum.org.uk under *are you there yet?*

Wolstenholme, A. and Constructing Excellence (2009) *Never Waste a Good Crisis* [Online] http://www.constructingexcellence.org.uk/pdf/Wolstenholme_Report_Oct_2009.pdf.

Chapter 11

BIM: A collaborative way of working

Philip King

Introduction

The Building Information Modelling (BIM) revolution is well under way. The UK Government has set a target to reach BIM Level 2 on public sector work by 2016, private sector clients are increasingly requesting BIM on their projects and the construction industry has been busy equipping itself with the systems and processes necessary to make it standard practice.

Ultimately, when all members of the professional and construction team have compatible technology and processes, BIM will completely change the way we work – for the better.

What is BIM?

There are a number of variants of BIM. For many years, architects and structural engineers have been developing their designs using 3D modelling or what we call 'lonely BIM', relating to specific parts of the project only.

The BIM being used today covers all aspects of a project from cradle to grave. It is a shared, digital representation of physical and functional characteristics of a built object, on which critical project decisions can be based.

There are some differences between the various levels. Level 2 BIM is a managed 3D environment held in separate discipline BIM tools with data attached. Commercial data are managed by resource planning software and integrated using proprietary interfaces or bespoke middleware. This level of BIM uses 4D construction sequencing and/or 5D cost information.

Level 3 BIM, on the other hand, is a fully integrated and collaborative process enabled by web services and compliant with emerging Industry Foundation Class (IFC) standards. This uses 4D construction sequencing, 5D cost information and 6D project lifecycle management information. For those that have yet to implement BIM at Level 2, you would be well advised to prepare for Level 3 at the same time to avoid further change in the future.

The following BIM maturity diagram, often referred to as the wedge diagram shows the progression from simple 2D CAD (Computer Aided Design) through to Level 3 BIM (Figure 11.1).

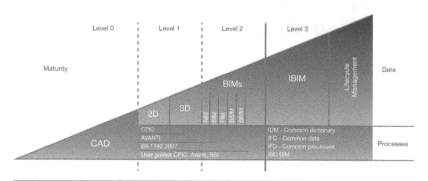

Figure 11.1 The wedge diagram. From Bew and Richards (2008)

BIM characteristics

- Digital
- Spatial (3D)
- Measureable/quantifiable
- Comprehensive – communicating design, performance, etc.
- Accessible to the whole team
- Durable

The benefits of BIM

BIM improves the way that the industry works together, making it more streamlined and efficient. Throughout a project, it leads to better collaboration, coordination and exchange of information.

The advantages for clients in both the public and private sectors are numerous, especially in terms of mitigating risk, reducing costs, minimising waste, shortening programme and smoothing handover.

BIM also has the potential to improve facilities management throughout the life of the building. The facilities manager receives high-quality as-built information at handover, integrated fit-out information and asset information in a single location. The BIM model, as new operational data is added, can highlight performance issues and energy spikes, allowing proactive and reactive maintenance to be greatly enhanced. Upgrades, churn and change during the life of the building can also be better managed. Gone are the days of bulky operation and maintenance manuals and performance statistics on pieces of paper – the new world is comprehensive, intuitive and immediate.

Outputs of BIM

1 Deliver design, 3D drawing, coordination and information sharing with the team
2 Create a BIM model that includes parametric information for the equipment
3 Use the architect's model for DTM, CFD, EPC, Part L, structures, fire and acoustics modelling
4 Carry out calculations such as heating and cooling loads, lighting, pipe, duct and cable sizing
5 Produce a schedule of rates and bill of quantities for procurement
6 Supply contractors' working drawings and 'builders' work in connections' drawings
7 Produce the asset register
8 Include operations and maintenance (O&M) information
9 Manage post-occupancy BREEAM assessments
10 Manage Part L metering information
11 Manage fit-out/churn

Making the most of BIM

In order to extract the greatest value from BIM, there are a number of steps that need to be taken. The first of these is internal.

For organisations adopting BIM, a cultural shift is needed. Working practices need to change and training is essential to use the system effectively.

The traditional way of working with other project team members needs to be replaced by an inter-relationship that enables design, costing and programming information to be brought together seamlessly in a single, centrally managed BIM model. Structural engineers, mechanical and electrical engineering (M&E) engineers, architects, contractors and specialists all need to know how the data they input impact on the 3D environment, which means having a more comprehensive understanding of each other's work.

Collaboration and the sharing of information are critical success factors for BIM. There are two crucial times when data exchange will have a huge impact on overall success:

1 At the award of the building contract, when the design model is assigned to the contractor, who takes responsibility for owning and delivering it.
2 On completion of the project, when the asset model, containing all of the built information necessary to operate the building, is handed back to the client by the contractor. Adopting the *Construction Operation*

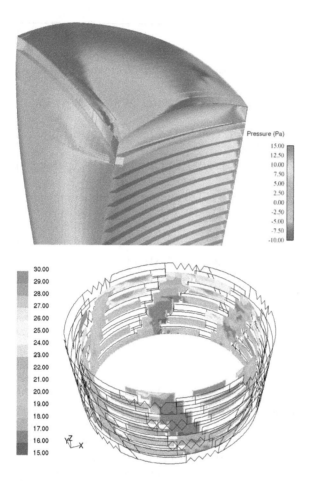

Figure 11.2 Thermal model

> *Building Information Exchange* (COBie) format is the recommended way of ensuring the information delivered back to the client is as useful as possible.

Contractors currently use a variety of 3D drawing packages to meet their requirements: estimating, direct output into ductwork fabrication and pipework prefabrication, procurement, etc. A single software package that achieves all of client's, designer's, contractor's and building operator's aims will become available to mitigate the risk of data being lost as they are transferred between different software packages.

For the client, there needs to be an agreed strategy for the use of BIM on the project. This needs to be clearly communicated and understood, so that

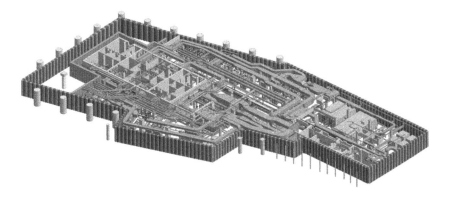

Figure 11.3 3D model

the project, building contract, delivery and occupation parameters are established from the start.

This leads to a BIM Execution plan including:

- Project objectives
- BIM objectives
- Implementation
- Project phases
- Collaboration
- Quality control
- Model structure and origin
- File naming and numbering

Deriving the real benefits from BIM

- All parties think in 3D
- Define ownership and transfer of the model during the project
- Change management
- Better integrated information within the model
- Greater consistency of platforms between consultants and contractors
- Improved data transfer between platforms
- Identify activities to hold within the model and those which should remain outside
- More 'BIM-friendly' contracts and appointments
- Asset and facilities management, measuring the real long-term benefit of BIM

To ensure quality, consistency and efficiency the day-to-day working, coordination and communication need to change. The following are some of the wider issues:

- Set up protocols correctly from the outset
- Commence 3D at the right time
- View progress information in 3D, not 2D
- Embrace new ways of working
- Agree design/coordination responsibilities
- Utilise an appropriate level of detail as the design develops
- Hold an appropriate level of information within the model at each stage

BIM and M&E engineering

Numerous opportunities exist to reduce the overall construction time and cost through the M&E services design. The most obvious are early contractor input, prefabrication, modularisation and early procurement.

As we have said, the success of BIM depends on information. Early involvement enables a better understanding of the project and its objectives, the sharing of information sooner so the 3D model can be created, and a smoother integration into the site and its services later.

This, in turn, means that a higher volume of prefabrication and modularisation is achievable within the timescales. Prefabricated modular construction can include plant rooms and plant housings, risers, distribution and meter cupboards through to kitchen and bathroom pods and assemblies. It can also include the installation and connection of complex systems, with modular wiring already well-established for activities such as lighting installations. Modularisation has a number of advantages including quality, reduced snagging time, speed of installation and economy, if volume can be produced.

M&E systems designers need to fully understand the construction sequence, particularly relating to basement and core construction, steelwork and cladding, in order to design the engineering systems in such a way that they integrate with the construction programme. This is particularly relevant to high rise buildings where the sequence of works in the core will generally be on the critical path.

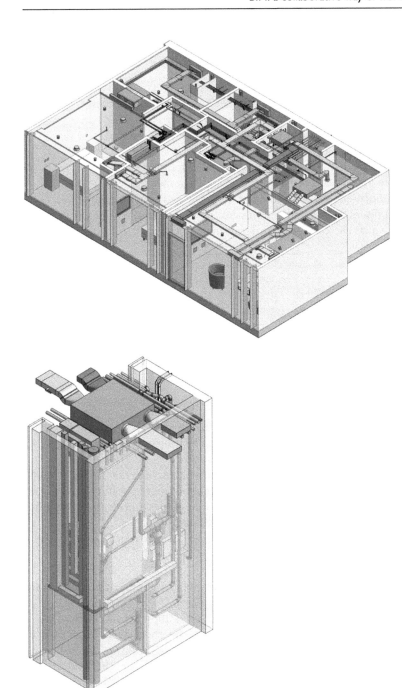

Figure 11.4 3D model – prefabrication

The steps to optimum M&E solutions

- A detailed and well-considered brief
- Early agreement on equipment and its procurement and, in particular, long lead items
- Timely appointment of sub-contractors and agreement of design responsibilities
- Effective risk management during the design stages
- Integrated design in a virtual 3D environment
- A thorough understanding of the proposed construction sequence
- Sectional testing and commissioning to minimise work carried out towards the end of the programme

---

Changing workflows

Coordination and integration between mechanical engineers, electrical engineers, public health engineers and BIM operatives mean a change in how we work together. Engineers need to think in 3D, plan and share detailed information to make the process as efficient as possible.

Calculations within the model

Within the BIM model, extensive calculations for sizing plant and distribution systems should be able to be carried out for heating and cooling loads, pipe, duct and cable sizing, fan and pump pressures.

CIBSE and BIM software developers are working together to improve the system and its functionality, which will enable more sophisticated calculations, multiple models on large projects and all design calculations within the model, reducing the need for additional packages.

The BIM impact

As with any new technology and process, some traditional approaches need to change and others need to evolve or improve.

There is a lot of information, and therefore liability, contained within a BIM model so appointments and building contracts need to change to encourage an open culture. Future contracts should cover the use of Level 3 BIM as well as outputs (2D drawings, specification, 3D model or all three), responsibilities, including the transfer of the model, and liability.

There should be a single owner of the 3D model during the design and construction stages so all parties contribute to one model, avoiding confusion

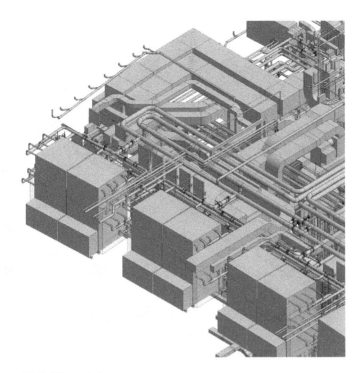

Figure 11.5 3D model

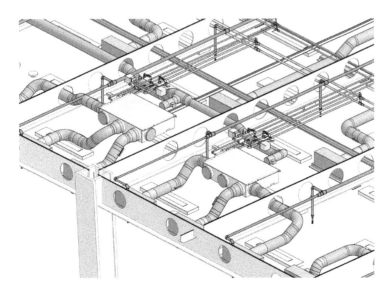

Figure 11.6 Ceiling void coordination

Figure 11.7 People and processes – not just technology

Figure 11.8 3D model – shell and core

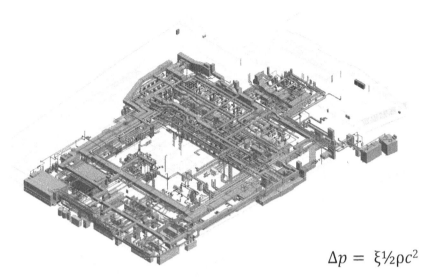

$$\Delta p = \xi \tfrac{1}{2} \rho c^2$$

Figure 11.9 3D model – distribution calculations

and delays. Additional responsibility and liability impacts would need to be reflected in the owner's contract.

The industry has an important role to play in the adoption and effective use of BIM. Here is what it needs to offer:

- A standardised library of parts
- BIM education – publications and courses so there is consistency in how we all use BIM
- Measurable and coordinated outputs
- BIM standards and codes
- Best practice examples – examining the potential and providing case studies of practical implementation
- A standard to enable CFD, DTM, EPC, Part L, calculations in the model
- Calculations for pump/fan sizing
- Agreed responsibility split between consultant and contractor to develop the installation details in accordance with the actual procured equipment
- A standard to create the bill of quantities and asset register, covering how all objects in the model are quantified, including small items such as fire dampers and sprinkler heads
- A process for tracking changes replacing traditional revision numbers
- Standardised schedules/templates linked to specification stage by stage
- Standard colours for each service, industry-wide
- A method for feeding plant/equipment quotes into BIM, taking into account variations in tenders

- A process for measuring and recording overall benefits of BIM: cost, risk and programme

On handover to the client, agreement on how the model will be updated after practical completion is needed. Fit-out information, churn, the Part L logbook and tenant fit-outs all have to be incorporated and an owner should be identified to manage this. To deliver the benefits of BIM in highlighting issues or areas for improvement, the model, O&M information and building management system must also all be linked.

Lastly, BIM fundamentally changes the way that we view costs on building projects.

Traditionally, the costs associated with each stage look like this:

Design	1 Unit of cost
Construction	10 Units of cost
In Occupation	100 Units of cost

Using BIM, they look more like this:

Design	1.5 Units of cost
Construction	8 Units of cost
In Occupation	90 Units of cost

With a change in cost structure should come a change in fee structures, with more time and effort put into the design stage. As an industry, we will need to be able to quantify money saved during construction and operation stages, thanks to early involvement and design.

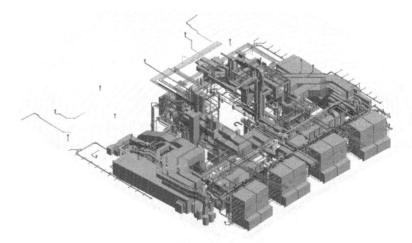

Figure 11.10 The model can be used for demonstration of space and access requirements, coordination, visualisation and prefabrication

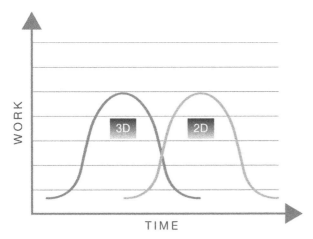

Figure 11.11 Industry experience

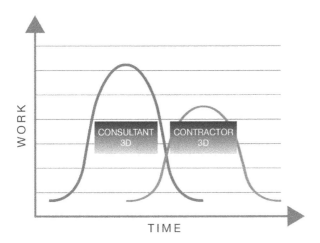

Figure 11.12 Actual experience

The future for BIM

BIM will become the standard way of designing and delivering building projects. The industry will become expert at operating within the BIM model and clients will embrace 3D over 2D.

We will also use BIM to do much more, including the creation of 3D Geometric models for the following:

• DTM – Dynamic thermal modelling
• CFD – Computational fluid dynamics

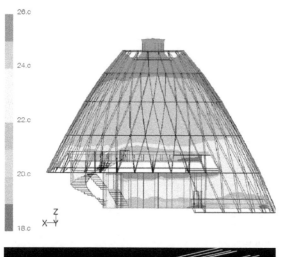

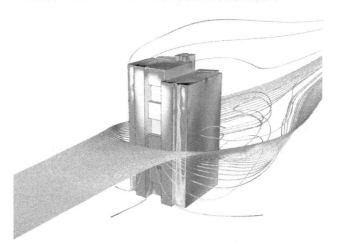

Figure 11.13 CFD model

- Comfort studies
- Pedestrian wind comfort analysis
- Natural ventilation studies
- Fire & smoke modelling
- Acoustics modelling
- Part L assessments
- EPC assessments
- Cost planning
- Programming

The success of BIM now and in the future depends on people, processes and the technology itself. If they are all working together, we will drive more efficiency into BIM, more value out of it and we will work as one team.

Chapter 12

Post-occupancy evaluation

Derek Clements-Croome

We need feedback from buildings in use in order to learn how to plan, design and operate better intelligent buildings. The RIBA launched a 2013 revision of the UK RIBA Plan of Work and Stage 7 covers POE (post-occupancy evaluation) (BSRIA E-News 18 June 2013). Clients also respect a good aftercare service. POE takes time and costs money but adds value over the long term. The methodology is likely to change as sensor technology advances to a point where occupants can self-monitor. The combination of physiological and psychological measures will allow judgements to be made which will reflect the state of well-being for an individual in a particular environmental setting. This will not only be valuable as anonymous personal but location-specific data for the facilities management system, but will help us to understand in a more holistic way the interaction between people and the building they inhabit. This could result in improved environmental criteria for various work settings.

POE is defined as the examination of the effectiveness of the design environment for human users. In contrast to an architectural critique which focuses on aesthetics, the evaluation of building systems or materials performance, a POE typically focuses on the assessment of user satisfaction and the sensory and functional fit with a specific space. Common complaints are that spaces feel stuffy, airless, too warm and lack personal control. POE can help to root out these problems for them to be dealt with. In the UK, maintaining temperatures of 24°C consumes about 15% more energy than controlling to 22°C. Moderately cooler temperatures such as these can be achieved by controlling heat sources (e.g. by using low-power lighting and computers), choice of materials, building mass and higher spaces, for example.

Many organisations now routinely include some form of pre-occupancy commissioning as part of the delivery of a project. The commissioning of building systems will involve verifying that the performance of a system meets the operational needs of the building within the capabilities of the design, as well as the owner's functional criteria. The commissioning process will include the preparation of the operator personnel (ASHRAE, 1995) and the setting of controls to the intended levels of performance. The commissioning process will also have to be documented.

The essential parameters monitored in POE are:

- the efficiency of building usage;
- the effectiveness of building usage;
- the performance of each building element;
- user satisfaction;
- the functioning of systems; and
- the utilisation of services.

While most aspects of the building services will remain relevant, changes will occur over time because:

- some problems will be phased out;
- new issues will come onto the agenda; and
- some of the technology will become obsolescent.

The quality of building services can be determined through indoor environmental variables such as temperature, indoor air quality, humidity, and light and sound levels. By using a wireless sensor network (WSN) to collect and then process data on indoor environmental variables, a dynamic picture of the actual state of the indoor environment can be produced. This can result in the efficient on-line control of systems, the increased well-being of occupants and decreased energy consumption.

The increasing miniaturisation of radio-frequency devices and micro electro-mechanical systems, as well as advances in wireless technology, has generated a great deal of interest in WSNs due to the fact that they can provide an infrastructure for gathering information about the physical world, including equipment and the behaviour of people.

A WSN is a group of wireless-linked sensors that perform distributed sensing tasks. WSNs have attracted great interest from academia and industry alike due to the diversity of possible applications. Recent advances in WSN technology have enabled the development of small, low-cost, low-power, multi-functional sensor nodes that allow communication over short distances.

By connecting a WSN to actuators in a building we can build up a wireless sensor actuator network. These not only collect information on indoor environmental variables, they also control the environmental systems. This creates a real-time, closed-loop control system in which the occupants are included.

The intelligent buildings research team at the University of Reading (Tamas *et al.*, 2007) is developing a WSN system and includes some of the following sensors:

- *accelerometers* to measure motion;
- *heat flux monitors* to measure how much heat is being given off by the wearer's body;
- *galvanic skin response monitors* to measure skin conductivity, showing the effects of physical exertion and emotional stimuli such as psychological stress; and
- *skin temperature monitors* to reveal how the body's core temperature is affected by physical exertion or the lack of it.

Other sensors becoming available include *ear* or *finger clips* which measure carbon dioxide partial pressure (pCO_2), oxygen saturation (SpO_2) and pulses; body *plasters* measure skin temperature, breathing activity, ECG heartbeat, blood pH and blood CO_2; *headsets* receive electric signals produced by the brain and monitor emotional state and facial expressions.

The *sense diary* (Croome, 1990; Mao *et al.*, 2007) can be used to record occupants' satisfaction and sensory well-being; the data collected allow facilities managers to improve the conditions for the occupants. The WSN includes temperature, humidity, light and air-pressure sensors that record the change in environmental parameters and then trigger actuators such as air conditioners and heating and cooling facilities. The diary can give a graphical representation of physical activity in terms of such variables as energy expended, metabolic rate and step count. It can also give a general indication of mood through the skin's galvanic response. Questionnaires may also be used to assess the occupants' state of well-being.

POE is recognized as necessary but how is it done and by whom? Methods are developing but BSRIA in their publication Soft Landings (BG 4 2009) encourage it to be integrated into the whole delivery process.

> Soft Landings is a framework that brings together best practice at all stages of a project. It has been developed to help clients, designers, builders, managers and end users to do this, by focusing the whole design, procurement, construction and commissioning process on improving performance in use.

As already mentioned it has been included in the 2013 version of the UK RIBA Plan of Work. Soft Landings complements rather than duplicates existing procedures used by the construction industry. Specifically, it has been designed to underpin environmental assessments such as BREEAM and LEED, all forms of energy performance certification, POE, building logbooks, green leases, and industry key performance indicators. It sits alongside any procurement process reinforcing five main areas:

1 Inception and briefing.
2 Managing expectations during design and delivery.

3   Preparing for handover.
4   Initial aftercare in the 4–6 weeks after handover.
5   POE meaning extended aftercare, monitoring and feedback over the first three years of occupancy.

Special issues of journals have been devoted to POE – see 'Making feedback and post-occupancy evaluation routine 2: Soft landings – involving design and building teams in improving performance', *Building Research and Information Special Issue: Building Performance Evaluation* (2007) Volume 33, Issue 4, pp. 353–360; and, currently, *Intelligent Buildings International* (2013).

Further reading

Mike Riley, Noora Kokkarinen, Michael Pitt, (2010) 'Assessing post occupancy evaluation in higher education facilities', *Journal of Facilities Management*, Vol. 8 Iss: 3, pp.202–213.

Chapter 13

Tenets for planning design and management of intelligent buildings

Derek Clements-Croome

We have defined intelligent buildings in terms of responsiveness to occupants; well-being of people; low resource consumption with low pollution and waste; flexibility and adaptability to deal with change. Above all that, intelligent buildings demonstrate an architecture that reflects the spirit of the age which future generations will enjoy and be proud of. Their development is along a continuum rooted in vernacular architecture and now moving with innovation towards buildings which are eco-effective; responsive to the occupants' varying needs; are healthy and simple to operate. Old and new buildings can share this evolution. Increasingly we observe how well the plant and animal worlds can show us economies in the optimum use of energy and materials in most beautiful ways, and this is leading to more examples of biomimetic architecture.

Intelligent buildings should be *eco-intelligent* and this means, in terms expressed by Goleman (2009), know your impacts; favour improvements; share what you learn. In this way buildings will be equitable for all in society; have long-life value; respect for nature. Wherever we build we have to fulfil human needs in an evolving technological world but set in particular cultural contexts. Braungart and McDonough (2009) believe form follows evolution rather than function, but in reality both apply.

These tenets are guidelines at this time but will change over time.

- Plan and design with an *integrated team* so that clients, consultants, contractors and facilities managers all develop a commitment to the project and want to fulfil the environmental, social and economic aims.
- *Systems and holistic thinking* are key.
- Assess the *impacts of the building* on occupants and communities nearby.
- *Occupants' behaviour* has a large effect on the consumption of energy and water so try to increase awareness of occupants to the impact of their actions on resources. Smart metering is a start but sensor technology is rapidly becoming applicable in building operation and for use by consumers of equipment. Data management systems are important to

give feedback on the performance of different spaces in the building. Use continual post-occupancy evaluation process.

- Use *whole-life value* or *whole-life performance* approaches to ensure quality as well as whole-life costs are taken into account.
- Aim for *simplicity* rather than complexity in operation.
- Think about *well-being and freshness* as well as comfort and consider all the senses and how air, view, daylight, sound, colour, greenery and space affect us in the workplace.
- *Connectivity* is important so there is interoperability not only between the systems and the building but also between the occupant and the building.
- Design for *flexibility and adaptability*.
- Think of an intelligent building as an *organism* responding to human and environmental needs but also one that needs to 'breathe' through the façade between the external and internal environments.
- The *façade* transfers light, solar radiation, air, noise and moisture, but also links occupants to the outside world so intelligent or smart façades allow these aspects to be controlled in a way which is functional but also enjoyable to those working and living inside the building.
- Plan the *facilities management* so the building and occupants are cared for.
- Balance *efficiency with effectiveness*. An air supply system, for example, can deliver the 'right' amount of air to a space and be deemed efficient but may not be effective in the space because it has no impact on the breathing zone where the people are.
- *Design beyond the expectations defined in Regulations.*
- *Keep abreast of the relevant fields of knowledge.*
- *Learn from other sectors and disciplines.*

Many companies today describe business intelligence in terms of being: smart to fulfil enterprise requirements and stimulate new insights; agile by having advanced integration which allows flexibility and adaptability; aligned so that pervasive intelligence links strategic, economic and operational management processes. So software products need to be innovative, agile and adaptable and this approach to business intelligence allows these aims to be achieved. Intelligent buildings, old and new, need this type of thinking throughout their whole life from concept planning to care in use and beyond.

Part II

Case studies

Chapter 14

Case study

Intelligent buildings in practice: the Buro Happold experience

Peter McDermott

Introduction

For many years Buro Happold has had an active international consultancy practice in the area of 'smart buildings'. The author has spent even more years working with specialist engineering teams, as both a design consultant and a systems integrator, to design, deliver and support integrated intelligent building systems in many and varied built environments. We have learnt from experience how best to design and deliver smart buildings; in a way that can meet the needs of both their owners and their users by using the best of modern technology and innovative engineering solutions. Over that time it has become increasingly apparent that there are several key success parameters in smart building developments.

Intelligent buildings must:

- keep the occupants safe and secure;
- optimise their living and working environments;
- minimise costs;
- promote productivity and increase profits;
- enable well-being and happiness;
- reduce energy and resource usage and greenhouse gas emission;
- enhance the surrounding neighbourhood;
- respond automatically to changing conditions;
- interact in an intuitive way with their users;
- use the best of modern innovative technology.

Integrated and intelligent buildings innovate in the use of technology but they are always the creation of integrated teams of intelligent people working together to achieve a common vision. This 'broad church' view of intelligent buildings means that we must seek to use the best available designers, construction methods and operational and facilities management techniques to optimise operation and whole-life value, not just the initial design and build phases.

As a foundation principle of Buro Happold we have always advocated and implemented a holistic multi-disciplinary approach to design in the built environment. Designing, building and operating modern, highly serviced or naturally ventilated buildings is a complex but rewarding process involving large teams of talented people with divergent skills, aspirations and styles of communication.

A foremost tenet of our firm's founder, Professor Ted Happold, was that the best results are obtained from a collaborative effort of the most diverse skills that can still successfully communicate and work together. We blend inspiration from the fields of architecture and engineering with the creative pragmatism of project management and practical understanding of time-served building professionals. This gives a result that is elegant and inspirational, efficient and cost effective to construct and a delight to live and work in. In these times of energy constraints and potentially catastrophic climate change any truly intelligent building must have sustainable construction and operation as fundamental drivers of design. Not to do so would be plain stupid.

The following are a number of anonymised case studies which demonstrate how we have applied the design of intelligent buildings on real projects.

Intelligent local government

Good examples of intelligent building can be found in the local government sector. During the economically buoyant first decade of the twenty-first century many local government authorities in Britain and Ireland and elsewhere invested in iconic new headquarters buildings. The first part of the process of creating intelligent buildings for these new facilities was to ensure that the design team at Buro Happold understood and worked with the council leadership teams and facilities managers. Design workshops were established to enable thorough dialogue between the building designers and the users who would inhabit the facilities after they had been delivered. The aim was to elucidate and document user needs and understand the whole life cycle business and future building operational requirements.

So where was the smart innovation here? We worked with leading architectural practices to develop and detail sustainable and architecturally striking solutions. We introduced new structured consultancy processes with key stakeholders in the council managements and workforces to highlight issues and crystallise the technology brief. As a result of understanding the user needs through this process we realised that the key factor for these projects was that any technology we were designing into the building needed to be fully integrated with the building management systems and support the organisational changes which were to be implemented during and through the move to new buildings and HQ facilities. Everything that we specified, installed and commissioned on the construction projects had to support the councils and their development of new operational cultures in new homes.

In some cases the new buildings that we were designing for our clients represented a quantum leap in style and complexity of accommodation. For instance, in one of the more interesting and challenging projects, the existing premises used by the council for their offices were situated in historic Georgian period residences. These wonderful buildings had been designed and built in a period when the state of the art for building services consisted of manually operated windows for ventilation, open fireplaces for heating and a very basic cold running water supply. The major 'Rocket science' used steam power and our modern technology of telecommunications and computerisation would most probably have raised cries of 'witchcraft!' Although they had been repeatedly refurbished and updated over the centuries, these fine Georgian terraced houses were a world away from the modern paradigm of open-plan and naturally ventilated offices and visually striking council chambers with which we planned to replace them. However, it should not be forgotten that, in its time, the Georgian sash window was a significant and radical, even 'smart', building engineering innovation. This radical window design allowed simultaneous top and bottom window opening which facilitated and promoted a convection-driven air circulation. This allowed the fresh air from outside to naturally ventilate the new style of tall but deep rooms. In a curious twist of fate the new buildings that Buro Happold designed to replace the Georgian town houses, included automatic controls on natural ventilation to induce cross flow ventilation across even deeper floor plates. We incorporated and monitored weather sensors via the integrated Building Management System (iBMS) to ensure that the ventilation openings were closed automatically when wind and rain exceeded comfortable limits.

Our structural design ensured that architecturally exposed concrete was used extensively. This created a large increase in thermal inertia which was used to smooth out more extreme fluctuations in outside air temperature. The holistic design ensures that the exposed concrete ceiling slabs of the building structure are exploited by overnight natural ventilation and free cooling when external ambient temperatures allow. This is used in the summer cooling season to reduce mechanical cooling load. Intelligent lighting systems were designed and installed to ensure that lights are only powered on when they are needed. Light activation and levels are automatically controlled by local light level and presence detection sensors, as well as programmable time schedules. In some of the smart building projects individual user interfaces are provided to allow occupants to control their local light levels via the iBMS.

Designing intelligent security and access control systems presented us with fairly typical engineering dilemmas. How were we as a design team going to meet statutory requirements to safeguard council employees but allow public access? Some of the staff worked in front line departments that dealt with difficult planning matters and parking and traffic offences. They had to be protected against potentially frustrated and aggressive members of the public. At the same time the councils were determined to maintain, and improve if possible, free access by constituents to their democratic representatives on the council. An integrated and zoned access control system was the engineering solution. This allowed different classes of users different access rights. It was configured and commissioned to reflect the careful public/ private demarcation of building spaces and took into account distribution of council departments throughout the building layout. As part of the holistic design the project team had to bear in mind the required operational, communication and supply adjacencies of the various spaces.

Formal and structured post-occupancy evaluation (POE) of actual building performance is vital if the development team want to obtain the best results from any new buildings. The key tool here for the researchers is a well-configured and easy to use iBMS. A careful analysis of real performance data allows the building engineers to fine tune the performance of the automatic controls once the building is occupied and real heating and cooling loads are in place. This ensures that the correct conditions are maintained, energy use is minimised and premature aging of equipment through oscillating and over responsive control is avoided. The initial data and analysis obtained in this process inform management action. POE will ensure that facilities support services and building users and occupants are fully trained in the most efficient operation of the new building. A bedding-in period is to be expected and where the complexity of the new building services is radically different from the old premises, as is the norm for these projects, a new lead facilities manager position needs to be created and filled by a suitably trained and experienced professional.

POE studies of buildings were carried out to provide data on the actual performance of buildings against the design intent. A key element of smart or intelligent design is measuring outcomes and using this information to inform subsequent design. Intelligent buildings are best achieved when design, construction and operation are viewed as an iterative process cycle resulting in continual and evolving design and incremental performance improvements.

Smarter cities and neighbourhoods

Many of our clients overseas are interested in creating complete smart city developments. The aims vary depending on the locations and particular objectives of the developers but, generally, there is a desire to create high-quality built environments and city neighbourhoods and districts that promote sustainability, leading-edge connectivity, innovation and creativity together with world-class business productivity. The use of ubiquitous information technology, distributed monitoring and control, and integration of facilities management oversight via the iBMS are recurring themes.

For example, in one major development in the Middle East one of our clients aims to regenerate the central city districts of a major and rapidly expanding city by recreating and re-visioning the traditional medieval medina in a modern mixed-use high-technology guise. We have worked with the client and whole design team over a long period to develop intelligent building systems that support comfortable living and productive work and also conserve valuable natural resources.

This particular city district is being developed in several phases but the master plan requires each building in each of the evolving development phases to connect back to the development's central facilities' and security management control rooms. Power and precious water resources are metered using smart meters to meet the US Green Building Council's Leadership in Energy and Environmental Design (LEED) energy efficiency certification and facilitate integration with the demand management systems of the local utility suppliers. All of these data are monitored, collated and transmitted to appropriate recipients for action by the iBMS.

The guiding aspiration is to provide smart homes for the residents and smart business accommodation in the mixed-use neighbourhoods. High-speed internet access and sophisticated controls combine with advanced audio-visual facilities to provide an ambience of effortless and luxurious living for the high-specification and technology-rich family homes.

Major challenges that we had to overcome were in defining the overall integration methodology and integrating the various network architectures with district-wide schemes, agreeing suitable compatible communication protocols and ensuring that the integrated intelligent systems provided an appropriate level of facilities management information in a suitable format in the locations required.

In another major project, this time in the Far East, our client's aspiration is to provide ultra modern and technology-laden environments in an iconic building designed for very wealthy individuals and families. This new building project is also one part of a much larger smart city development. The aims for this whole smart city are that ubiquitous interconnected microcomputers will saturate the city environment and built environments informing, automating and empowering the lives of the residents and visitors. In this way the iBMS becomes more deeply embedded in the building and infrastructure.

Our intelligent building design had to reflect the interconnectivity requirements of the smart city master plan and distinct services as well as the immediate architectural vision of the particular residential building that we were designing.

The result is a comprehensive IT infrastructure which forms the basis of an iBMS. The iBMS supports the most modern personal technology needs and desires of the wealthy residents with unobtrusive security measures and automatic and integrated building controls that minimise energy and resource use.

Intelligent airports: information and control in airport terminals

Buro Happold has carried out numerous design and development projects with several major airport operators around the globe. As well as the more traditional engineering disciplines of structural and building services engineering, a modern airport needs many specialist and more esoteric disciplines to work together in an integrated endeavour.

For instance, we needed to model people flows. This helps optimise the trade-off between robust security measures, which are unfortunately so necessary in our modern world, and the operator's strong desire to enhance passenger experience and to optimise the available leisure opportunities, especially last minute shopping!

Safe egress from the terminal buildings in case of fire or other emergency must be balanced against the need to keep airside and landside activities fully separated and prevent unauthorised access by the sightseeing public or potential criminals.

The airport operational systems must be designed to assist passengers to find their way in a stressful situation in a potentially bewildering environment. Way finding and passenger information screens are integrated with information and control systems for ticketing, flight despatching and baggage handling, etc. On most modern airport developments we work with clients to develop integrated airport systems. These use common databases and integrated network communications to provide the real-time information environment needed to support the operational management of the airport.

Airports, more than most buildings, need to perform at peak levels for large parts of day and night. Time for maintenance is limited and intelligent building management systems that minimise manual work and oversight are very welcome. Airport operators generally have well-defined standards of the control and monitoring that they require and an intelligent design must incorporate these and innovate using the latest in technology. Our engineers worked with many stakeholders to ensure that the systems were designed and integrated to achieve the highest levels of availability and usability. Integration between the airport business systems and those that automate the building services provide the best result. A common database is designed to consolidate all the data and information and present it in the most appropriate format to the many different types of users. The integration of the airport building and business systems requires a dedicated master systems integrator to coordinate all the constituent systems and providers. Here, the iBMS is integrated into even more complex and wide-reaching airport and airline information systems.

Optimised people movement is critical to the timely departure of flights and prevention of overcrowding in the various lounges and departure gates. All the processes, from parking, drop off, checking in, baggage drop, security clearance, shopping and boarding, need to be performed seamlessly. Huge investment in the technology has seen rapid advances in the automation of many of these processes and the systems integration behind them is now even more mission critical.

Intelligent offices: smarter ways of working and more productive workspaces

In recent years we have seen big changes in the way people work and the places they choose to work in. Rapid change in information technology not only changes means of communication from post and fax to email, instant messaging and video conferencing. Within our own design business, which operates across the world in many time zones, we have adopted new methods of group collaboration and information transmittal. All of these changes are impacting on the design of workspaces.

Modern offices support creative workshopping and interaction spaces, areas for quiet, intense, solitary intellectual work, market and dealing arenas and more informal spaces for social and casual interaction. Clever design is needed to get the right mix and flexibility of these spaces. Inspirational environments are vital for supporting well-being and creativity so vital in the modern knowledge economy. Comfort and security must be delivered without fuss or complication to working patterns. Once again the engineers and technologists need to work closely with the client and architectural colleagues to deliver a holistic response.

We have helped deliver Smart Office solutions around the world for many years. One case study project involved the creation of a speculative office development in central London with strong green environmental credentials but the flexibility of office space design to meet the potential needs of a wide variety of potential clients. We designed the shell and core building services to achieve a BRE Environmental Assessment Method (BREEAM) Good rating for the initial building. iBMS fan coil unit and lighting controls were implemented in the initial Cat A fit out. This provided the flexible floor requirements needed by the developer and their property agents for this competitive market place of prospective tenants.

In this case, the eventual occupant was a major local governmental organisation that had a strong corporate aspiration to fit out the building to even more exacting and challenging environmental criteria. We worked with the client to realise their aspirations for flexible and productive workspaces containing open-plan offices, meeting rooms and specialised control rooms. In the process we modified the existing systems and designed and integrated significant further sustainable technological innovations including combined heat and power (CHP), fuel cell, absorption refrigeration, wind turbines and on-floor demand-led ventilation control requirements. The building design achieved a BREEAM Excellent rating. All of these systems were monitored and controlled by an open protocol BMS employing web server technology to deliver targeted information to different user groups.

Further work with the client team was carried out on this project after handover and occupation to ensure that the complex and innovative technology was delivered and integrated into the ongoing operations of the overall organisation.

Intelligent project completion and handover

The construction phase of a project is a complex process involving numerous teams of tradesmen and technical specialists working together in challenging circumstances to tight deadlines. When the building is completed these teams disperse to new challenges. A new team of facilities management staff moves in and takes over. Sometimes this can result in temporary confusion and all the knowledge of the constructors is lost. In the UK, legislation ensures that information for the safe operation of the building must be handed over in the form of a comprehensive safety file. To get the best from the investment in the building, as much knowledge as possible must be captured and shared.

In one case we designed a large building campus facility providing broad-cast and headquarters accommodation for a very large media organisation. We were tasked with leading the construction project team in a thorough process of knowledge transfer to the incoming facilities and operations teams. The buildings had a highly resilient infrastructure to support business critical broadcasting. These systems had to be complex with multiple failure and backup response scenarios to be understood. A systematic and layered process of education and knowledge transfer began early on in the design and construction process.

We produced tailor-made documentation for the main systems and organ-ised interactive training workshops for the facilities staff. This knowledge transfer was supported and reinforced by full attendance at commissioning stages to pass on as much knowledge as possible to the incoming operators both formally and informally.

Ensuring a full 'soft landing' at the time when the users take over a completed construction project is a key aspect of the delivery and operation of an intelligent building. Where thorough training and handover processes are implemented and good operating and maintenance documentation is delivered, the chance of a facility operating efficiently and productively, as it was designed to, is vastly increased. Just as with a high-performance car, building performance will decrease with time unless the systems are periodi-cally re-tuned. Automated fault detection and diagnosis (FDD) programmes can help prevent the gradual degradation of performance and highlight problems. These techniques help target the FM maintenance efforts where they are most needed and where the financial returns are the greatest.

Intelligent energy performance

Given the critical nature and growing impact of climate change, all intelligent buildings must conserve the use of energy and resources and, in particular, reduce emissions of greenhouse gases.

This can be most easily achieved by careful implementation of direct digital controls (DDC) using energy-efficient control strategies for individual plant items with a carefully crafted graphical interface to provide information to the users. Together these form an iBMS which is the intelligent nervous system and brain of an intelligent building.

On a recent very large commercial office project, once more in the Middle East, we designed a robust, resilient, high-capacity communications network connecting an integrated building management and control system. Tasked with providing high-capacity communication to the internet we designed communication networks utilising CAT 6 copper and fibre optical communications networks and compatible open standard communication protocols to transmit and integrate the data to and from the many intelligent building systems. Once normalised, collated and stored these data can be manipulated by sophisticated and complex software algorithms to spot patterns and identify automatically where equipment is performing less than optimally and raise alarms and maintenance work orders. Early versions of such automatic FDD software have now been implemented and have been demonstrated to reduce energy consumption and maintenance costs in a financially significant way.

A key success factor for buildings that reduce carbon emissions is the ability to measure performance automatically and present information to users that enables them to act and operate the buildings in the most energy-efficient manner. This is best achieved using an iBMS.

For the whole life and future operation of the building these iBMSs will facilitate the transfer of large amounts of building energy and usage data for further manipulation and processing to provide actionable information for owners and occupiers.

We have discovered, in case after case, that

- elegant integrated building systems and technology,
- operating with enlightened and informed users,
- to provide delightful, sustainable, efficient and productive built environments

define the intelligent building in practice.

Chapter 15

Case study

Post-occupancy evaluation: University of Nottingham Jubilee Campus

Mark Worall

This case study analyses the design, operation and occupant satisfaction of three buildings, the International House, the Amenities Building and the Sir Colin Campbell Building, that form part of the Jubilee Campus expansion at the University of Nottingham, UK, in terms of the definition of 'intelligent buildings' as those that

> are responsive to the occupant's needs, satisfy the aims of an organisation and meet the long term aspirations of society, sustainable in terms of energy and water consumption and able to maintain a minimal impact on the environment in terms of emissions and waste. They are healthy in terms of well-being for the people living and working in them and are functional according to the user needs.
>
> (Clements-Croome, 2007: 55)

The Jubilee Campus at the University of Nottingham, UK, is a modern purpose-built campus which extends to 65 acres and is located only one mile from the main campus at University Park. The initial phase was opened in 1999. The Jubilee Campus has been developed on a site that was previously of industrial use, and has attempted to demonstrate the value and environmental benefit of brownfield regeneration.

Since its initial development, the Jubilee Campus has been undergoing a major expansion of its facilities including the commissioning of new and iconic buildings. In the first stage of the expansion (2005–2008), three

Table 15.1 Project team

Client	University of Nottingham
Architect	Make Architects, London
Project management	Gardiner Theobold LPP, London
Building Services	AECOM Building services, London
Structural Engineering	Adams Cara Taylor, London
Contractor	Rok Sol Construction Ltd, Nottingham

buildings comprising the International House, Amenities Building, and Sir Colin Campbell Building were designed and built at the Jubilee Campus. The criterion for the delivery of a successful design was primarily to create signature buildings of iconic form that would promote the business of the University and also provide the right mix of accommodation. A key element would be to encourage innovation in building design, promote sustainability and energy efficiency.

Brief descriptions of the three buildings are given below.

The Amenities Building is a four-storey building of $2,607\,m^2$ useful floor area and has its main façade facing Triumph Road with a main entrance at both ends of the building. The ground floor comprises teaching/seminar rooms arranged around a central core of toilets, staircase/lift and service risers. The ground floor also includes a café (Café Aspire) and student support area. The teaching/seminar rooms continue on the first floor which also provides accommodation for the Graduate Centre. Perimeter rooms make full use of the heights where possible and provide good natural light. On the upper floors six studio flats and two three-bedroom flats provide accommodation for University staff and visitors.

International House is a five-storey building of $3,190\,m^2$ useful floor area and provides predominantly office accommodation and is configured in a similar form to the Amenities Building around a central service core which includes some internal teaching rooms. The main entrance lobby and exhibition space is accessed via Triumph Road with a further entrance facing the campus.

The concept design of the Amenities Building and International House is encapsulated in the following statement from Make Architects Ltd.

> A series of pavilion structures set within the landscape, which grow from the ground like geological land forms. The finished buildings thus represent an abstraction of this initial design concept, being wedge shaped emerging from the ground plane to address Triumph Road.
>
> The unique form of each building is given additional visual impact by a rainscreen cladding system of terracotta tiles which are arranged in a random pattern of rich reds and browns, representing a contemporary take on the City's traditional red brick architecture.

The Sir Colin Campbell Building is a three-storey building of $4,181\,m^2$ useful floor area and contains a range of facilities which include office, event, exhibition and teaching spaces and incubator units for start-up businesses. The majority of these facilities are concentrated in the portion of the building

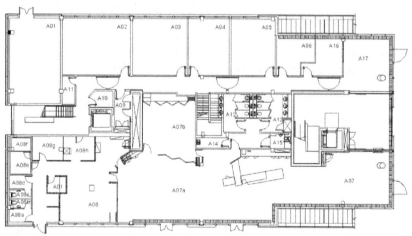

Figure 15.1 Photograph of Amenities Building and floor plans of floors A, B, C and D

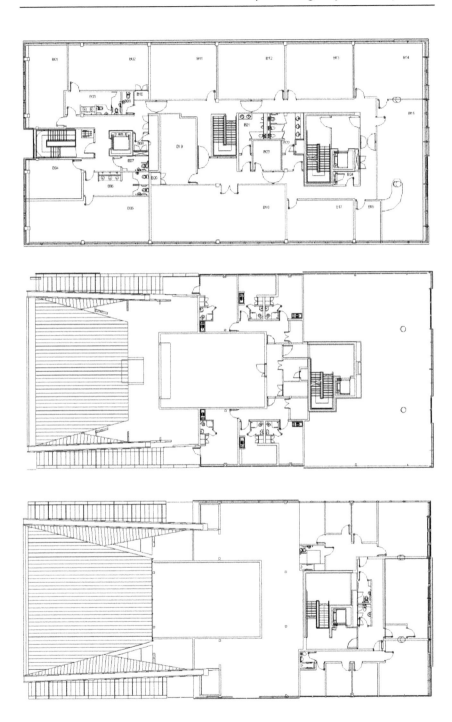

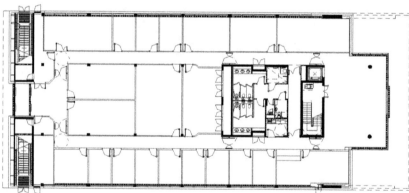

Figure 15.2 Photograph of International House and floor plans of floors A, B, C and D

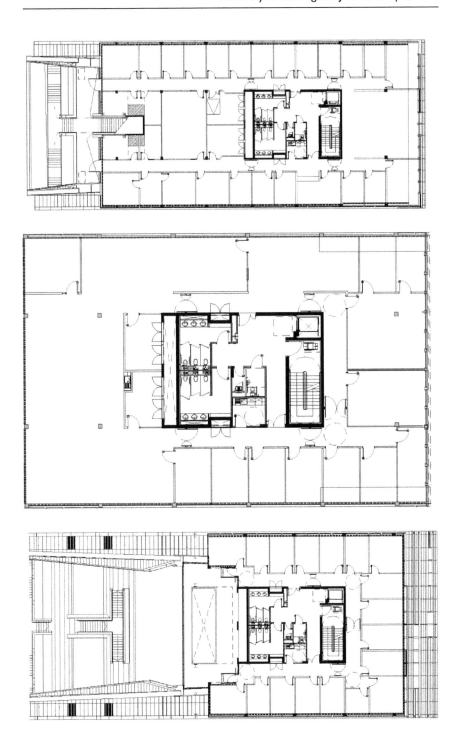

to the west of Triumph Road, while the more elongated structure on the other side of the road, to the east, houses the business incubator units.

The concept design of the Sir Colin Campbell Building can be summarised by Make Architects as:

A form that emerges from the ground plane merging with the surrounding landscape in the shape of grass mounds rising either side of Triumph Road to bridge the primary route through the campus and create a new gateway.

The Sir Colin Campbell Building is clad in zinc shingles to tie in with the visible zinc standing seam roofing of the Amenities Building and International House, but contrasting strongly with their terracotta cladding.

The three buildings were completed in 2008 with all academic departments able to move into the buildings on a phased basis for the start of the next academic year.

Post-occupancy evaluations (POE) were carried out by QTC Projects Ltd in 2010/2011 comprising questionnaires, interviews and a workshop. Detailed POE reports of this building complex can be found on-line at the University of Nottingham (2011).

Passive and active environmental quality control

Chapter 8 described how a combination of passive and active technologies and design strategies could be adopted to minimise the consumption of energy while providing a productive and pleasant environment in which to work. The buildings in this development minimised the use of energy by specifying high thermal insulation standards exceeding minimum building regulation specifications, using internal thermal mass to store and release heat slowly and to reduce peak heating and cooling through night cooling, controlling solar gain through solar control glazing, encouraging natural daylight through well positioned windows and, where possible, encouraging open-plan spaces (where natural light could not penetrate a space, light-pipes were installed), installing motion-controlled lighting, utilising renewable and low-carbon heat by exploiting the embodied energy stored in a nearby lake, encouraging natural ventilation through low-level entry points at the front of the buildings and high-level exhaust points toward the back of the buildings, integrating windcatcher type natural ventilation systems to ventilate interior spaces not well served by standard natural ventilation, developing lakes that provided natural environments for a wide variety of wildlife as well as for rainwater collection and storm water attenuation and for a heating/cooling source. As well as designing for a minimal impact on

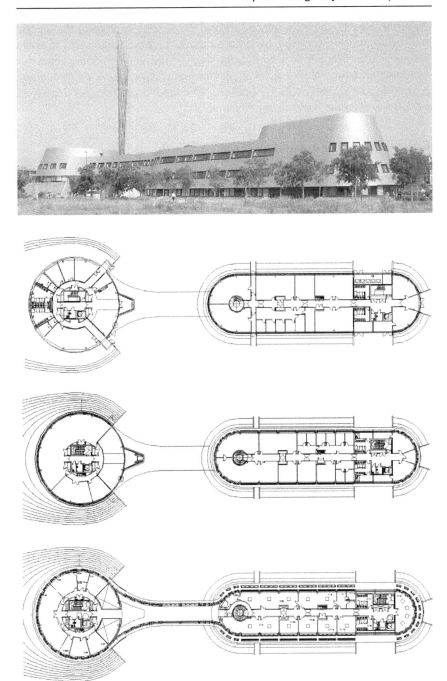

Figure 15.3 Photograph of Sir Colin Campbell Building and floor plans of floors A, B and C

the environment, the buildings addressed the aim of improving the health, well-being and productivity of building users. This was addressed through design by encouraging natural lighting and ventilation, providing openable windows and control of the local environment, designing for flexible and adaptable workspaces and developing natural landscaped features to provide a connection to the natural environment.

Building performance evaluation

Lake source reversible heat pump for heating and cooling

One of the innovative features of the development was to use renewable and low-carbon resources to provide heating and cooling. Advantage was taken of a nearby lake already in existence on Jubilee Campus, which could act as a heat source in winter and a heat sink in summer in a reversible heat pump system. The heat pump system provided heating and cooling to air-handling units within the building and fresh air was delivered by pressurised floor plenums. Stale air was then extracted through grilles above doorways and ducted back to the air-handling units where heat could be recovered. Thermal mass was provided internally with exposed concrete columns to regulate temperatures and provide night time cooling.

Winter heating

Problems were encountered in operation of the heat pump system in winter and, overall, 66% of users who responded to the questionnaires were dissatisfied with the level of heating in winter. In analysing the problems, it was observed that quality control was not monitored at the design stage, some of the assumptions made during the design were inadequate, such as assuming constant lake temperature when circulation in the lake produced variable temperatures, neglecting the effect of a weir connecting a nearby river to the lake, and finding that the depth of the lake was inadequate to provide the energy reserve required to meet the heating demand of the buildings. The Operations and Facilities section of the Estate Office at the University of Nottingham, who were responsible for the operation and maintenance of the building after handover, considered the system to be 'overcomplicated, inefficient and with no back-up'. In its operation, it was found to be unreliable, which was mainly attributed to some control hardware components and complex BMS software.

In responding to the problems, a number of modifications were adopted but without solving the problem of poor winter heating performance. In the end, gas-fired boilers was installed to provide back-up to the heat pump system in key parts of the building.

Temporary electric heaters were used in some cases when temperatures became uncomfortable, but had only a limited effect as their distribution had to be restricted due to the electrical load on the buildings, particularly International House. The report found that it was important that the Estate Office has a contingency plan in place that will ensure business continuity and procedures established clearly setting out how this should be handled. Adequate information issued through the Estate Office Helpdesk is crucial.

Summer cooling

Summer temperatures within the building have been much more satisfactory, but it has been reported that some of the rooms were felt to be too hot and stuffy, especially the internal rooms and rooms with no opening windows.

Sound insulation

In all three buildings, it was found that there was inadequate sound insulation between rooms, resulting in excessive noise transmission across rooms and partitions within rooms.

Lighting

Lighting, both natural and artificial, was considered generally good. The configuration of multiple windows in many of the rooms in the Amenities Building and International House gives a good level of light and adds to the quality of the rooms. There have been some comments concerning glare from some of the high-level windows in the offices.

Recommendations from POE

- The Services Consultants should continue to be retained on future projects on the client side and have experience of ground/lake source heat pump systems and undertake a quality and monitoring role as well as more involvement in the system design and specification.
- Any innovative system needs to incorporate some form of back-up in the event of failure of the main plant.
- The checks and tests made at the commissioning stage need to be further strengthened and the system modelled on winter time temperatures and conditions prior to acceptance at handover.
- The appointment within the contract of a specialist commissioning engineer may be appropriate on future capital projects.
- Low temperatures on Monday mornings have caused discomfort during cold spells. Where the temperature drops considerably out of normal hours, the building management system should be adjusted to compensate for the lower temperatures.
- Operations and Facilities should meet with user representatives to discuss

the issue of communication and how this can be handled better in the future, should a similar situation arise.

- The Estate Office should have a contingency plan in place that will ensure business continuity is sustained and procedures established clearly setting out how this should be handled.
- Ensure adequate information is issued through the Estate Office Helpdesk.
- The rooms identified as having complaints about hot and stuffy conditions in summer need to be inspected and the temperatures and ventilation tested/re-balanced.
- To remedy the complaints of noise it was recommended that construction issue drawings should be checked to confirm whether any additional insulation had been included. Partition junctions need to be checked at ceiling level and below the raised floor where acoustic baffles may have become dislodged. Further discussions with users should take place to identify specific rooms where problems have been experienced and in-situ sound insulation tests undertaken.
- To reduce problems of glare in some high-level offices, blinds have been fitted to windows, but this is not a consistent policy and further assessment is recommended.

Energy consumption and carbon dioxide emissions

The Energy Performance of Buildings Directive (EPBD, 2002) requires that the energy performance of all new buildings other than dwellings be evaluated with a standard calculation methodology that complies with the Directive. The Simplified Building Energy Model (SBEM) (BRE, 2012) is a computer programme that evaluates annual energy consumption and carbon dioxide emissions of a building, enables its performance to be compared to a reference building and to be rated using an asset rating (AR), as defined below:

$$AR = 50? \frac{BER}{SER}$$

where BER is the actual building emissions rate and SER is the standard emissions rate of the reference building. An energy performance certificate (EPC) can then be issued in compliance with the EPBD.

Carbon dioxide emissions were predicted at the design stage using SBEM. Emissions were then calculated based on the actual buildings using SBEM in 2009 and EPCs were awarded. Table 15.2 describes the predicted emissions, the AR and calculated emissions based on the actual buildings.

SBEM evaluation showed that carbon dioxide emissions were higher than predicted at the design stage. In the Amenities Building and International House, the ratings meet the benchmark criteria for the reference building type, but in the Sir Colin Campbell Building, the rating is better than the

Table 15.2 Predicted and actual annual carbon dioxide emissions of case study buildings based on SBEM rating

Building	Predicted CO_2 emissions per unit area (kg CO_2/m^2)	Predicted CO_2 emissions (kg CO_2)	AR (Benchmark AR)	BER (kg CO_2/m^2)	CO_2 emissions (kg CO_2)
Amenities Building	87	226,809	73(73)	107	278,949
International House	53	169,070	71(71)	84	267,960
Sir Colin Campbell Building	47	196,507	58(72)	101	422,281
Total		592,386			969,190

benchmark. Overall, the EPC rating shows that carbon dioxide emissions are about 60 per cent higher than the predicted values.

Actual electricity and gas consumption has been measured since the building was opened and metered values for 2011 are shown in Table 15.3, together with carbon dioxide emissions. This shows that actual carbon dioxide emissions were within about 5 per cent of the values calculated from the SBEM programme and overestimated emissions slightly. Actual carbon dioxide emissions were about 1.5 times higher than the values predicted at the design stage. Emissions are higher than predicted mainly because gas-fired back-up boilers were installed in all three buildings to compensate for inadequate performance of the lake source heat pump system in winter and because of excessive leakage from ductwork.

Table 15.3 Actual annual energy consumption and carbon dioxide emissions of case study buildings

Building	Gas consumption (kWh)	CO_2 emissions (kg CO_2)	Electricity consumption (kWh)	CO_2 emissions (kg CO_2)	Total CO_2 emissions (kg CO_2)
Amenities	94,279*	17,274*	485,490	253,243	570,092
International House			574,846	299,575	
Sir Colin Campbell Building#	103,730	19,005	617,173	321,633	340,638
Total	129,009	36,279	1,677,509	874,451	910,730

* Gas metering shared between buildings
# metering errors March to December
Carbon factor electricity = 0.52115
Carbon factor gas = 0.18322

User/occupant satisfaction

Chapter 11 describes the importance of POEs in delivering a successful building in terms of performance and user/occupant satisfaction. The POEs found that the iconic forms did not impact negatively on the functioning of the buildings apart from difficulties in integrating standard furniture layouts in irregularly shaped envelopes.

The use of non-load-bearing internal partitions was used to provide a flexible and adaptable space, and floor boxes in raised floors which incorporated service voids for power and data conduits, enabled building use to be flexible. However, some internal rooms did not have direct access to natural light and these spaces were not well liked. The floor boxes were not well liked due to perceived difficulty of access from under the desk for portable equipment.

One of the positive aspects of these buildings was the quality of the natural light and its effect on mood and well-being in open teaching rooms and open-plan office spaces. Dissatisfaction with rooms was found mainly in International House, which comprises mainly cellular office accommodation and internal teaching rooms without access to natural daylight. International House was occupied by the International Office, the Centre for Education Language Education (CELE), the Institute of Work, Health and Organisations (IWHO) and the School of Contemporary Chinese studies. At the briefing and early design stages, the International Office played an important role in defining how they wanted their space to be configured within International House. This has also helped in the design of their reception/waiting area, whereas CELE and IWHO found the reception areas did not function as originally intended and the absence of any waiting space was very noticeable. In addition, their responses to shared office accommodation were generally negative. An important aspect that emerged from the POE was that occupants who had been involved from the design stage found the spaces generally well liked, whereas occupants who were not consulted found the spaces less adequate and more dissatisfaction was found.

In the Amenities Building, favourable reports were received on the use of the Postgraduate Centre with good use made of the dedicated seminar room and postgraduate lounge, which is used both as a study area and as a social space.

The residential units in the Amenities Building are arranged on the top two floors. The units, which form part of the University's staff accommodation portfolio, are very popular and there is a quick turn round to maximise letting potential which is usually for one to three months' duration. The accommodation is used for new starters, staff relocating and visitors.

A well furnished communal lounge is also provided and, combined with the use of Café Aspire creates an environment conducive to meeting colleagues or other people. The three-bedroom units work well as they can be let to a family or a group of colleagues. Again, the Housing Co-ordinator advised on the mix of units at the design stage and this early involvement has contributed to the success of the units. Overall the facility works well and should be replicated in other buildings where residential units for staff and visitors are to be provided.

The fit out of the Café Aspire in the Amenities Building has worked well, demonstrated by the level of use of this facility. The success of the venue has put pressure on the bookable private dining area especially at lunch time and a separate dining room would have been preferred. However, due to the layout of the building and internal circulation this would be difficult to retrofit.

Conclusions

This case study has presented a summary of the results of POEs carried out for three buildings on Jubilee Campus – the International House, Amenities Building and Sir Colin Campbell Building – in terms of the definition of an 'intelligent building'. In most cases the buildings have performed well, are generally well liked by the occupants and users and meet the operational needs of the organisation. However, design issues involving innovative heating and cooling technology resulted in dissatisfaction by a majority of users during the winter heating season. Additional back-up was needed from gas-fired boilers, impacting negatively on the original concept and its potential for reducing carbon emissions. Encouragingly, the conclusions drawn were not to dismiss innovative solutions, but to ensure quality control and adequate back- up at the design stage. It also emerged that staff and users of the buildings who were involved in the early stages were more likely to be satisfied with their accommodation. Although it may be difficult to consult with all staff that might use the buildings, it is important that time and effort is made to involve as many of the potential occupants as possible at the early design stage.

Chapter 16

Case study

Tall building design and sustainable urbanism: London as a crucible

Ziona Strelitz

Tall buildings offer important scope for sustainable urbanization through their architectural and urban design potential, but they are also resource intensive and inclined to negative impacts. Recognizing these helps to inform and promote sustainable urban development. The prospect is promising. Significant advances in energy-saving and modelling technologies can shift tall buildings towards more sustainable design, operation and social impact. The scope is signified by new designs for London, where the combination of heritage controls, sustainability focus, high land values and appetite to build tall has created a *de facto* 'design laboratory'. London's conditions have galvanized design strategies that target a wide range of sustainability credentials, encompassing: energy consumption, use of renewables and embodied energy; occupier benefits – including environmental comfort, spatial quality and flexible internal planning; and positive urban conditions in terms of accessibility, microclimate, visual permeability and public realm. In combination, these establish a wider range of success factors than energy efficiency alone, supplementing the widespread global foci on height metrics and commercial viability with a more holistic tall building design agenda. Despite relatively modest heights in international terms, London's new exemplars offer strategies to inform more sustainable design, wherever tall buildings are developed.

Keywords: social impact; sustainable; tall buildings; urban planning.

Tall buildings: Environmental potential

The theoretical sustainability benefits of tall buildings are well recognized. These are urban compaction, through scope to accommodate significant density on limited footprints, and their support for efficient mass travel when tall buildings are clustered at public transport nodes (Newman and Kenworthy, 1999). Further, green potential relates to their scope for mixed use. This affords energy transfer between distinct uses that vary in their respective profiles of demand over the 24-h cycle, and reduced travel through users' proximity to a range of facilities close at hand. Tall buildings also offer scope for social benefits – functional and aesthetic – for their occupants, their more occasional users and people beyond their precincts.

Inherent demands and tensions

Nevertheless, height challenges sustainability in key respects relative to low buildings of equivalent area (Strelitz, 2005), detracting from their positive potentials. Strategies to mitigate these characteristics are therefore relevant to enhance tall buildings' green performance. The factors relate to space efficiency, wall-to-floor ratios, wind loading and structure, vertical circulation, thermal management, lighting and logistics. A suite of reciprocal tensions is involved:

- The ratio of net to gross space decreases with added height, as the service core must increase in response to structural needs and the requirements to access and service additional floors. While larger floors make for more area-efficient tall buildings, the deeper space provided requires more mechanical and electrical servicing to afford occupier comfort.
- Wall-to-floor ratios are inherently high in tall buildings, involving more embodied energy to provide an equivalent quota of area. This runs counter to lean design. The issue is accentuated with shallow floor depths, but slender buildings dispose more internal space to reliance on natural light, reducing artificial lighting and cooling loads.
- Added building height increases wind load, requiring enhanced stiffness. This involves more energy in the building's production. This embodied energy becomes proportionately more significant as tall buildings become more energy efficient operationally.
- Tall buildings' dependency on vertical movement systems impacts on core volume and energy use, generating further tensions in the sustainability equation. Technological solutions to achieve efficient lifting and core compaction – for example, by means of double-decker lifts – may be at odds with user aspirations for simple, legible systems. They may conflict with Yeang's call for low-key provision to facilitate pedestrian movement in tall buildings: without complex infrastructure, and offering increased journey variety (Yeang, 2002).
- Most of the energy used over a building's life is on lighting and thermal management. While large floor plates are more demanding in these respects, they offer better area efficiency.
- Tall buildings are most relevant in densely developed settings. However, these are the conditions in which close proximity to other buildings calls for building form to respond to heritage considerations and/or rights of light. It is also where complex substructures may be needed to accommodate obstructions from adjacent infrastructure.

- Constructing tall buildings poses inherent challenges to safe and efficient movement of people, materials and equipment. These logistical requirements are accentuated in congested urban conditions – precisely where tall buildings are most applicable. This is addressable through well conceived, integrated design and construction strategies to facilitate the delivery of tall buildings with minimal disruption to surrounding urban life.

London as a crucible of relevant design innovation

London has played an instructive role in shaping sustainable tall building responses. The drivers are its stringent planning context, and high land values. In combination, these have catalyzed responsive design and strategies that offer thinking with wider geographic relevance (Strelitz, 2006).

London's historic narrative is legible in its numerous commercial, civic, cultural and religious buildings, among which St Paul's Cathedral has been vested with strong pre-eminence. This status is supported by the planning requirement for open viewing corridors from key vantage points towards St Paul's to be maintained. An intricate raft of further planning constraints relates to other heritage buildings, medieval streets, variety of plot size and shape, rights of way and light, and highly diverse land ownership (Figures 16.1 and 16.2).

Figure 16.1 Aerial view of City of London conveying its complexity

Image credit: Kathleen Tyler Conklin. www.flickr.com/photos/ktylerconk/.

This matrix of constraints is associated with policy caution that arose in large measure from negative reaction to London's first generation of tall buildings that were developed in the 1960 – 70s. Considerable disaffection with these concerned negative social aspects of their use – including the inappropriate accommodation of families with young children at high levels, a lack of social cohesion and a sense of anomie. However, antipathy was also engendered by the buildings' physical effects. These included hostile impact on the public realm, difficult access into and around them, and unpleasant microclimate impacts like windiness. Another negative factor was weak economics relative to building usage – with inefficient ratios of core to typically small floor plates.

The resultant cautious stance was revised following the subsequent development of Canary Wharf. With little by way of existing street pattern or other constraints, tall construction in these former London docklands proceeded relatively unfettered. These buildings, with their large orthogonal floor plates and extensive floor area, provide the 'big space' that facilitates the interior planning that large corporate organizations welcome. This promoted Canary Wharf as a competitive location to the City for London's financial and professional services occupiers. The City then developed a more facilitative approach to tall buildings, alongside its continued custodianship of heritage and urban character. These latter attributes are deemed integral

Figure 16.2 Framework of view controls in City of London relating to St Paul's Cathedral

Image credit: KPF. Architect: KPF.

Figure 16.3 Competition to the City of London: vertically extruded large orthogonal footprints at Canary Wharf

Image credit: Peter Pearson. www.flickr.com/photos/peterpearson/.

to London's international appeal and have remained pivotal in its planning (London Planning Advisory Committee, 1998; Greater London Authority, 2009) (Figure 16.3).

Meanwhile, London's government assumed leadership in adopting and evolving the sustainability agenda. Its policy was to promote development that 'thinks lean, thinks clean and thinks green'. This objective to promote the development of greener buildings provided another significant planning influence. In parallel, a focus on responsible corporate governance has motivated occupiers to adopt and demonstrate positive practices in their real estate selection.

Although their average 60 storey-height contrasts with the super high-rise buildings in the global tall building arena, London's new tall building designs have had to meet this significant array of planning requirements. They are therefore instructive in informing sustainable architecture and urban design further afield.

London's exemplars: more sustainable performance systems

Sophistication in the design process supports strategies that reduce the requirement of added stiffness required by height to have a heavier frame.

The structural stability solution for The Pinnacle, to be built at Crosby Court, 22– 24 Bishopsgate in the City of London, is provided by a frame around the building perimeter – essentially using columns, beams and diagonal bracing. The advanced computational modelling and iterative design involved in generating this have enabled elements of structure to be removed where it is not needed, and reinforcement where extra stiffness is required. The result is leaner in materials use. Given the large number of floors in tall buildings, such savings that can be achieved by avoiding redundancy are magnified (Figure 16.4).

The process is similar to that adopted in developing the exterior diagrid structure for the CCTV building in Beijing. This too involved considerable reduction in the number and size of the structural steel members, with testing to demonstrate that the process compromised neither stability nor robustness. A separate, but related, result for both these buildings is their aesthetic impact

Figure 16.4 The Pinnacle's elevational treatment showing the frame around the building

Image credit: Eamonn O'Mahony. Architect: KPF.

in the public realm. In both cases, the varied articulation of structure is expressed in a more randomized external appearance.

Irrespective of the potential for regenerative power solutions, numerous strategies are now available to achieve efficient lifting, enabling a reduction in elevator numbers and an increase in building population. The repertoire includes a range of options for use in varying combinations: sky lobbies served by express shuttle elevators, zoning of floors served by local elevators from transfer levels, double-deck cars, twin lifts operating in a single shaft, destination hall call on both double-deck and conventional elevators (with new algorithms modelled on passenger behaviour and expectations). The scope that these offer for core compaction and intensification of building use relative to embodied resources enhances tall buildings' sustainable potential.

The new vertical access strategies also afford social benefits, based on a more humanistic tall building ecology. Heron Tower, approaching completion at 110 Bishopsgate in the City of London, incorporates a vertical series of 'office villages' surrounding triple height north-facing atria that offer fine views and day lighting, volumetric variety and a community focus. The design affords scope for stairs to interlink the set of floors grouped around each atrium. This option can be installed up the height of the building, increasing user choice in their mode of internal circulation. The arrangement facilitates vertical pedestrian movement and a sense of community, promoting sustainable living in the building. Facing outward, the glazed elevators offer stimulating city views for their passengers and animation of the external elevation for passers-by, enriching urban experience within the building and beyond (Figures 16.5 and 16.6).

Figure 16.5 Heron Tower: good daylight, sense of community and scope for easily accessible stairs within an office village

Image credit: KPF. Architect: KPF.

The Pinnacle's design harnesses dedicated and shared express and local lifts, with scope for destination hall call to operate at peak times. The strategy is supported by sky lobbies that can also serve as 'social hearts' at different levels in the building. These spaces are poised to offer spatial variety and afford appropriate settings for sociability and collaborative activity. The mix will be further extended by the public viewing gallery on top of the building. Knock-out panels provide potential for inter-communicating stairs.

Figure 16.6 Heron Tower: animating the City through its vertical circulation

Image credit: Hayes Davidson. Architect: KPF.

Minimizing non-renewable energy in building performance is the most common sustainable design objective. Relevant tall building designs attuned to London's local conditions, and using passive controls and sustainable materials and systems, include: The Leadenhall Building (The 'Cheesegrater'), under construction at 122 Leadenhall Street, London Bridge Tower (The 'Shard'), under construction at 32 London Bridge Street, Southwark, 30 St Mary Axe, (The 'Gherkin'), in the City of London, The Pinnacle and Heron Tower. Active multi-layered façades are integral to the design strategy in all these cases: intercepting solar gain, controlling heat gain and loss, minimizing reliance on artificial lighting, and drawing in air from the exterior to facilitate natural ventilation. In London's moderate climate this enables parts of these tall buildings to function without air conditioning at appropriate times of the year. Important to evolving aspirations for building use that are evidenced by post-occupancy research, these strategies also offer enhanced user control and more active mediation with external conditions – experiences that building users now increasingly seek (Strelitz, 2008).

Heron Tower exemplifies the positive scope to meet these objectives. Its critical starting point – appropriate building orientation and the protective south core that shields the office floors from solar gain – are integral to the overall design. Daylight admitted to the floor plates reduces reliance on artificial light, and the building skin facilitates natural ventilation. The interactive façade is triple glazed – with a single-glazed outer pane, double-glazed inner pane and cavity mounted blind for shading. The low iron glass provides high transparency to afford excellent light and views, reduce energy use, and enhance user experience by contributing to spatial quality and the sense of connection to the external realm.

Other elements include the photovoltaic (PV) installation on the core, high-efficiency plant for heat recovery and low-energy cooling systems. The design also deploys low-energy lighting and lighting control, low-energy cooling and decentralized air plant – to limit service to occupied floors and save energy use on unoccupied floors. The projected carbon emission is 38 per cent less than the standard required by the Building Regulations (Part L2, 2002), excluding the energy demanded by the building's occupants, for example, through use of information technology (IT) (Figure 16.7).

ZZA's post-occupancy research identifies increasing user concern for a direct relationship to the outdoors, as well as scope to open windows (ZZA, 2009, 2010). The Pinnacle's triple-glazed façade is designed to meet these interests in enabling users to open windows throughout the year. The strategy is conceived to offer thermal management centred on natural ventilation in moderate weather – approximately half the year in London. The design has been supported by testing to ensure that differential pressures on opposing elevations will promote air flow through the building. Situated externally to the thermally sealed glazing units, the sun shading will keep radiant solar heat outside the building, with its position between panes – and thus protected from wind behind the outer pane, facilitating its consistent use.

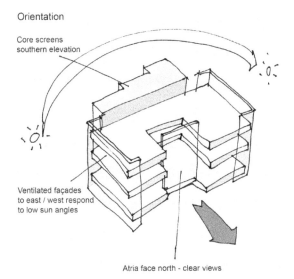

Orientation

Core screens
southern elevation

Ventilated façades
to east / west respond
to low sun angles

Atria face north - clear views

Figure 16.7 The bedrock of Heron Tower's environmental design: appropriate building orientation

Image credit: KPF. Architect: KPF.

The triple-glazed cladding modules will be fabricated off site and delivered already assembled, for installation from the given floor on which they are being positioned. Efficient construction such as these processes contribute to sustainability.

High-performance envelopes also exploit passive solar heating and heating from internal gains – notably from IT and artificial lighting. This strategy can allow mechanical heating to be reduced to a level that is just complementary. Harnessing the heat generated by building use can obviate the need for additional heating in tall office buildings, except to pre-heat space in colder weather. While the full potential of this strategy relates to moderate climates, the principle can be applied to other climates as local scope for passive building servicing allows. The potential for aggregate energy savings magnifies the relevance of mixed mode strategies across new building stock.

The exemplars show the scope for tall building design to draw from a repertoire of energy strategies according to context and site conditions. These represent a suite of potential opportunities. The functional uses and mix to be accommodated are also pertinent. A profile of building uses such as The Shard's – retail, residential, office and hotel – offers inherent scope for energy transfer as shown by the early environmental strategy shown in Figure 16.9. The strategy has evolved over time and has not been implemented as shown.

No building can be fully reliant on natural light, especially in the 24-h global business culture most typical in big city environment where tall

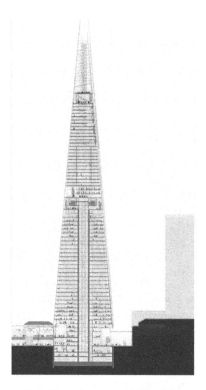

Figure 16.8 Mix of uses in different sections of The Shard, and 'social hearts' at ground, mid-and upper levels

Image credit: Renzo Piano Building Workshop. Architect: Renzo Piano Building Workshop.

buildings have most relevance. This calls for energy-efficient lighting in all internal spaces, including high-performance fixtures with central controls in public spaces, light-emitting diode (LED)-illuminated exit signage and occupancy sensors in unoccupied areas like stairwells.

Considerable hope for sustainable building operations has been vested in the generation of renewable energy. Within the overall potential for this, tall buildings' offer distinctive scope, based on height affording a clear sunlight path to PV panels on roofs or upper portions of façades. The associated technology can have additional expressive value in promoting sustainability. The visible statement made by the large PV array on Heron Tower's south core is an example.

Given the relative immaturity of the systems in use at this stage, guidance on their appropriateness and selection is helpful. The proposals for Heron Tower benefited from 'route maps' to evaluate the potential renewable energy technologies relative to a given building's specifics (Faber Maunsell, 2004).

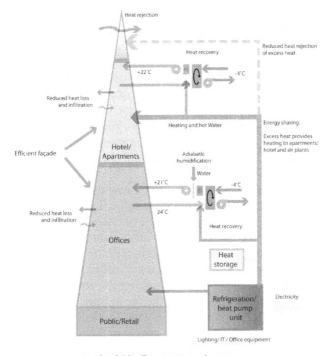

London Bridge Tower, proposed system

Figure 16.9 Scope for energy sharing with The Shard's mixed use design
Image credit: Arup. Architect: Renzo Piano Building Workshop.

In this case, wind turbines, solar thermal systems, biomass heating, biomass combined heat and power (CHP) and ground source heating and cooling were judged non-applicable.

Cities offer additional scope for sustainable use of locally generated energy, as the use of energy close to its point of generation minimizes loss in transmission (Faber Maunsell, 2002). CHP, deploying waste to produce power, heat and cooling in useable form, with reduced CO_2 emissions and overall thermal efficiencies of up to 80 per cent, is especially efficient to source energy in densely occupied settings.

Lifecycle performance and building adaptability

The significant number of floors in tall buildings increases the benefits from using materials with lower environmental impact. A life cycle analysis to optimize design at concept stage can promote potential savings by design optimization. An early life cycle analysis of alternative floor plate structures for the Beetham Tower at Holloway Circus in Birmingham compared the environmental performance of concrete and steel systems. Using BRE's 'Ecopoints method' (BRE, 2004), it identified pre-cast concrete plank floors on steel beams as lowest in embodied 'ecopoints', enabling effective integration with the services, and flexibility for potential change in use.

Design for occupational flexibility is important to promote sustainable building use, with shorter-and longer-term horizons both relevant. The Pinnacle's predominant intended use as a commercial office building is buffered by space planning tests to ensure its floor plates' scope to accommodate the respective layouts favoured by insurance, legal and financial organizations – sectors whose physical locational clusters intersect in the vicinity of the site.

Longer-term, re-using buildings for alternative purposes as occupiers' needs evolve optimizes both their materials and usage value. London also demonstrates effective approaches to the adaptive re-use of tall buildings rather than their demolition. Numbers of first-generation office structures, such as City Point at 1 Ropemaker Street in the City of London, are now successfully converted to provide a contemporary standard of office accommodation. Other examples are of adaptation for alternative uses, for example, high-quality apartments – as at The Panoramic, the former British Gas building at 152 Grosvenor Road in Pimlico, and students' residences – as at Nido King's Cross, the former Nat West Property offices in Pentonville Road.

Implications for urbanism

London's demanding planning environment and high land values have catalyzed more sustainable buildings with strongly modelled forms. Their urbanistic impact is significant. The 'Gherkin's' prominent role in promoting London's successful bid for the 2012 Olympic Games, and in civic and commercial imagery since, exemplifies the potential for distinctively formed tall buildings to become icons. The Shard's strong form has similar resonance. In an age where visual communication is now so central in all walks of life, building image is powerful currency for developers, occupiers and cities (Figure 16.10).

Given their scale and visibility, tall buildings' form and orientation can have a dramatic impact on the urban prospect, both positively and negatively. This is recognized in the planning strategy for London, althrough its provision for 'tall building clusters' at certain agreed locations (Greater London Authority, 2004). Intelligent modelling can help to optimize urban aesthetics through the provision for vistas and views – distant, middle and local, as well as enhancing functional aspects of the public realm. Modelling tests and simulations to identify negative environmental impacts such as overshadowing and wind can support sustainable tall building design, by averting uncomfortable conditions. As tolerances to discomfort vary according to the impacts, functional uses and time of day, specificity is important, with due attention to the proposed building use.

Grimshaw's unbuilt proposals for the 217 m 43 storey Minerva Tower, on the site at 138 – 139 Houndsditch at the eastern gateway of the City that has since been occupied by The St Botolph Building, remain instructive. The elegant 'bookend' design was for 100,000 m2 of net allowable office space plus retail and restaurant facilities, with public access at the top of the building and notable public realm improvements at its base. The skilful arrangement of form and volume responded to a highly complex suite of planning constraints based on heritage views (Figures 16.11–16.14).

Form, massing and core positioning are key to the urban views and vistas that a tall building can allow or occlude. Modelling is a valuable aid in

Figure 16.10 Strong visual impact: The Shard's strongly modelled form

Image credit: RPBW – photo Hayes Davidson and John Image credit: Grimshaw Architects LLP/Smooth. Architect: Mclean. Architect: Renzo Piano Building Workshop. Grimshaw Architects LLP.

Figure 16.11 Elegant response to view constraints: The Minerva Tower

Image credit: Grimshaw Architects LLP/Smooth. Architect: Grimshaw Architects LLP.

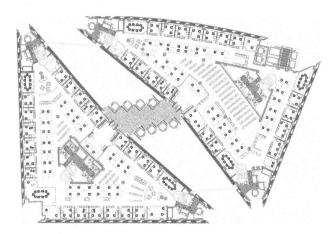

Figure 16.12 Floorplan for The Minerva Tower, incorporating range of depths

Image credit: Grimshaw Architects LLP. Architect: Grimshaw Architects LLP.

Figure 16.13 Daylight, external aspect and fine spatial quality between the
blocks: The Minerva Tower

Image credit: Grimshaw Architects LLP/Smooth. Architect: Grimshaw Architects LLP.

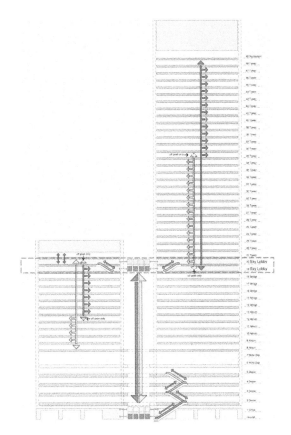

Figure 16.14 Vertical transportation scheme for The Minerva Tower,
incorporating sky lobbies

Image credit: Grimshaw Architects LLP. Architect: Grimshaw Architects LLP.

achieving the City of London's twin objectives to maintain cherished views without frustrating commercially viable development. The forms of The Pinnacle and The Leadenhall Building both involve some loss of floor plate to accommodate strategic views, while incorporating compensatory height and bulk where this is visually acceptable (Figure 16.15).

The Pinnacle's design unifies the growing cluster of the City's tall buildings and facilitates visual permeability across the cityscape, as does the design for The Leadenhall Building. Compared to more uniform towers, the variation in both buildings' forms and elevations should enrich urban experience by offering a change in aspect as one's eye moves up and around them. Heron Tower already provides this interest and aide to orientation in the City (Figures 16.16 and 16.17).

Figure 16.15 The Leadenhall Building: optimizing floorplate and visual permeability across the City

Image credit: Rogers Stirk Harbour •Partners/CityScape. Architect: Rogers Stirk Harbour •Partners.

Figure 16.16 Unifying and reinforcing the City's tall building cluster: The Pinnacle

Image credit: CityScape. Architect: KPF.

Figure 16.17 Unifying and reinforcing an urban cluster: The Pinnacle
Image credit: Eamonn O'Mahony. Architect: KPF.

The generation of good public realm on the ground plane associated with London's current tall building design exemplars is precedented by earlier innovation – notably at Manhattan's Rockefeller Center and Foster's HSBC in Hong Kong. The Pinnacle's design provides for enhancement of its surroundings, with trees lining the street, improvements to an adjacent square, improved access for mobility impaired people to negotiate a change of levels, and by acting as a canopy to calms conditions in what was an unpleasantly windy locale.

Tall buildings' scope to provide public realm encompasses internal as well as external spaces. The Pinnacle's design incorporates ground floor retail facilities that are approached from the street, and a passage through the building honouring an historic public route. This will be signified externally by an apex in the building canopy and a vertical gap up the elevation, on axis with the entrance to the passage. The feature is designed to be visible from afar and expressed as a column of light at night, promoting orientation in the public realm (Figure 16.18).

The tapered southern profile of the 47-storey, 224 m-high Leadenhall Building inclines away from and protects the view through to St Paul's Cathedral. The base of the building incorporates a seven-storey-high land-scaped open space, with bar, restaurant terraces at levels two and three. This will provide for free public access, including an area for events and performances, and space for public use without the requirement to be a customer.

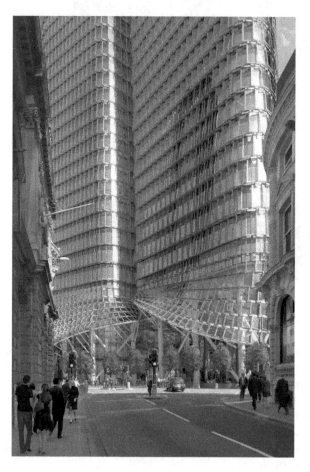

Figure 16.18 Promoting legibility: The Pinnacle's vertical gap on axis with the
entrance

Image credit: CityScape. Architect: KPF.

While physically enclosed, this design of this large area affords a high level of permeability, visual and physical, with a pedestrian route connecting Leadenhall Market and Lloyd's of London to the south with St Helen's churchyard and Bishopsgate to the north.

Design strategies such as these that provide enhanced permeability, connectivity and legibility help make cities walkable, safe and productive, contributing to sustainable urban living.

Enhancing building occupant experience and meeting other imperatives

Recent decades have witnessed an increasing occupier and user focus on functional quality. Inhabitants look for buildings to cater for evolving work and lifestyle modes and cultural preferences. Strong 21st-century themes for workplace buildings are variable depth to accommodate both individual and team working, and a high environmental standard that offers users external aspect, natural light and preferably a degree of local control (ZZA, 2002). London's new tall building exemplars are showing the potential to meet this agenda, with strategies that draw on the principal constituent elements – core, structure, storey height, floor plate, vertical access, façade and mechanical systems. Each system impinges on the others in shaping the experience of building users, as well as meeting the challenging requirements of building form that are critical from the perspectives of sustainability, commercial viability and – in a highly curated urban setting like London – of planning.

Using intelligent design to widen tall buildings' sustainable benefits

London's new tall building exemplars benefit from advanced modelling and analysis tools to develop designs that are more sustainable in energy use as well as urbanistically. Building identity and image are important design objectives for cities as well as developers and users. Given their visibility and the investment involved, this is especially so for tall buildings. The strong aesthetic impact and expressive value of London's well crafted new tall building designs responds to the distinctive challenges and opportunities poised by the city's development context. For building occupants, the enhancement of spatial quality, servicing and circulation offered by the exemplar designs represent quality-of-life enhancements inside the buildings. In parallel, the public realm improvements associated with and leveraged from intelligent and thoughtful tall building development widen the constituency of beneficiaries to everyone using the city.

Acknowledgements

Thanks to colleagues from the following organizations who have kindly discussed ideas and shared material relating to the development of this article: Arup, British Land, Foster and Partners, Grimshaw Architects LLP, Kohn Pedersen Fox Associates (International) London, OMA, PLP Architecture, Renzo Piano Building Workshop and Rogers Stirk Harbour + Partners.

Previously published in *Intelligent Buildings International*, 3(4): 250–268 (2011)

Chapter 17

Case study

Druk White Lotus School, Ladakh, India

Francesca Galeazzi

Introduction – Ladakh

Sometimes known as 'Little Tiber', Ladakh is an ancient kingdom located high in the Indian Himalayas, close to Tibet's western border. This remote high altitude desert (3500m above sea level) is an environment of extremes, surrounded by the magnificent mountain ranges of the Great Himalayas and the Karakoram (reaching over 8800m) and cut off by snow for around six months of the year, with winter temperatures dropping as low as −30°C. In the summer, temperatures rise to 40°C due to the intense solar radiation, and the snowmelt brings the rich fertile valleys alive. There is minimal precipitation all year round; the little water available for use comes from the snowmelt of the glaciers that is collected by the River Indus.

Ladakh is one of the few remaining mountain societies where a traditional Tibetan Buddhist way of life is practised. It is a remote community, isolated during the winter due to the closing of the only road that connects it with the other Indian states. Nevertheless it is a community under an immense pressure to change, principally due to rapid social changes, to the impact of 'Western lifestyles' especially on younger generations, tourism, technology, and overall associated with being part of modern and fast-changing India.

Background to the school project

Local communities, committed to having their children educated in their own language and culture, requested help from their spiritual leader, His Holiness the 12th Gyalwang Drukpa.

In 1992 the Drukpa trust was formed and registered as a charity in England. His Holiness the 14th Dalai Lama is the patron. The aim of the Drukpa Trust is to raise funds to design and build a new school and with the donation of the land by the villages of Shey, the construction of a school started, the Druk White Lotus School, a model for appropriate sustainable education and development.

Since 1995 a team of architects and engineers from Arup and Arup Associates has worked on the project on a mostly voluntary basis and is

Figure 17.1 Ladakh, a high altitude desert

Figure 17.2 Ladakh, a high altitude desert

© Arup Associates

Figure 17.3 Ladakh; traditional villages are still closely linked to monasteries and a rural economy

© Christian Richters

responsible for the masterplan, concept and detailed design for the whole school complex. The plan for the school includes a phased building programme for about 800 students, including 200 residential students from remote areas, ranging in age from 3 to 18 years. Designed to provide education from kindergarten and nursery through to secondary education and vocational training, it includes a range of facilities including classrooms, workshops, computer facilities, science laboratories and studios, a library, sports facilities, residential accommodation, dining hall and support facilities.

In 2001 the nursery and infant block was completed, followed by the phased construction of more school buildings and residential accommodation. The idea behind the phased construction has to do with the difficulties of building in such an extreme environment, and also to allow the school to grow, almost organically, to accommodate the growing students' population. Currently about 60% of all planned buildings has been completed.

Design philosophy: why is it an intelligent building?

The Druk White Lotus School is conceived as an entirely sustainable project. This includes the architectural and engineering design, as well as the construction, operation and educational mission of the school – all of which aspire to a modern education in harmony with local traditions, culture and religion. The ecological context is fragile, so the site strategy aims to ensure a completely self-regulating system of water, energy and waste management.

The application of Arup's engineering skills and Arup Associates' architectural expertise in sustainability has been intentionally directed to an innovative 'clever low-tech' design approach. The design team strongly believed that building in a context like Ladakh would need to be responsive to that particular extraordinary environment; buildings would need to be

Figure 17.4 Opening ceremony of the first school block in 2001. View of the internal courtyard and the north building of the primary school

© Caroline Sohie

Figure 17.5 Local stone and timber building, with fully glazed south-facing façade for passive solar heating

© Christian Richters

designed as intuitive and easy to operate; the site would use the available resources in a sustainable and appropriate way; the overall design would support the teaching and learning activities by providing simple, flexible and comfortable spaces that celebrate local culture and skills. This was in clear contrast with the high-tech approach often adopted in the area in the recent years, especially by foreign firms, which has generated buildings that are difficult to use and require vast skilled maintenance, with the consequence that in time, those buildings stopped working and their performance decayed rapidly. Arup's team believed that innovation was not in loading the buildings with technology and mechanical systems, but in developing simple passive systems that could be easily communicated to the end-users and intuitively operated.

The design and construction of the school buildings is greatly reliant on local tradition. The UK design team went through a challenging learning process to appreciate local materials and construction methods, and to adopt them innovatively into the building design. Strategies included:

- using locally available materials, which have the least impact on the environment
- exploiting natural ventilation and passive solar heating
- minimizing energy use and emissions
- minimizing water use
- refining and adapting traditional techniques to provide modern solutions.

In terms of technical input, Arup has developed and uses powerful software tools that allow accurate analysis of issues such as the design of passive solar systems, the feasibility of using wool or paper as insulating material, ventilation, daylight or the use of double glazing. In addition the team had access to the firm's broader experience in seismic design and appropriate design for developing countries.

The very different cultural backgrounds that separate designers working in London, Tokyo, New York and Ladakh, particularly with regard to local perceptions of appropriateness and feasibility, has always been a key challenge. For example, infrastructure and services are generally not seen as important in Ladakh, as local buildings rarely have running water or a reliable power supply. At the same time, the design teams had to recognize different expectations with regard to completeness and finish, as well as the operation and level of maintenance of the buildings.

This process of assimilation and adjustment was not always easy. It involved continuous simplification of construction details and operational elements; the preparation of clear operational manuals and, most importantly, working closely with the site and local school teams. Training teachers and students is playing a crucial role in securing the correct operation of the buildings, and ultimately, in the success of the school.

Masterplan

The school is in the village of Shey, 15km from the main town Leh, close to the river Indus and its surrounding irrigated fields, and on a gently south-sloping desert site surrounded by two important monasteries.

The masterplan can be divided into four main areas: the first, the site entrance, caretaker house, visitor centre and bus drop-off from the road to the south, gives pedestrian access to the second, the daytime teaching areas, and the third, the residential spine rising to the north. The fourth area, comprising the water and energy infrastructure and the sports facilities, is located separately alongside a service track to the west.

The plan for the school buildings is based on the traditional nine-square grid of the *mandala*, a symbolic figure of particular significance in Buddhist philosophy, surrounded by a series of concentric circles formed by low walls, *stupas* and willow trees. At the heart of the *mandala*, the circular library building forms the centre of the plan and offers an open air temple and assembly space.

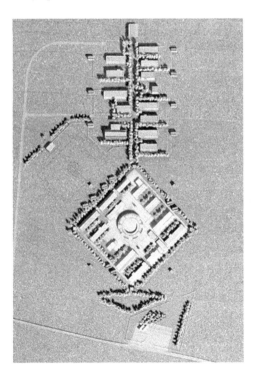

Figure 17.6 Model of the masterplan. The school buildings are arranged within the rotated square of the *mandala*, while the residential spine unfolds along the north–south axis

© Arup Associates

The masterplan takes maximum advantage of the site with the primarily one-storey buildings orientated differently according to their function. The *mandala* is oriented 30° towards the south-east so that all buildings within it can gain maximum benefit from the morning sun, which at this altitude is abundant even in the coldest winters. Direct sunlight can in fact reach 1000W/m² at 9 am on a clear December day!

The plan for the residences is oriented along a north–south axis, with all residential buildings facing south, thus exploiting the beneficial solar radiation throughout the day.

Materials and resources

The construction materials – stone, mud mortar, mud bricks, timber and grass – are mostly indigenous to Ladakh, with careful auditing of the sustainable resource supply. The stone for the walls is actually found on the site and the mud for mortar, bricks and roofing is excavated nearby. Timber is grown locally wherever possible: poplar and willow used are provided from nearby monastery plantations and local producers. Ladakh lacks the natural resources required for more complex construction, so glass, structural timber, cement and steel have to be bought from outside the local community and delivered from other parts of Kashmir or India. Generally, the use of imported products has been radically minimized.

Figure 17.7 Local materials are used: stone, mud mortar, mud bricks, timber and grass. The project seeks to reinterpret and celebrate local construction techniques and traditional architectural details

© Arup Associates, Caroline Sohie, Francesca Galeazzi

This contrasts with the trend amongst local architects and engineers to design new buildings in steel and concrete, an approach often wrongly perceived as 'better', because it is identified with a Western modernity. These buildings are often poorly designed and constructed, with the result that they deteriorate more rapidly. Imported materials have a far higher embodied energy than the traditional stone, mud and timber construction, which has proved itself successful over time.

Furthermore, the project seeks to include and celebrate the use of local expertise in craftsmanship, traditional construction skills and the symbolic aspects of the architecture, reinterpreting local traditional techniques with a contemporary pioneering and sensitive approach in all aspects of the design. For example, the solid granite walls used for all wall structure are formed and finished by skilled masons from stone from the surrounding boulder field, the use of valuable soil is kept to a minimum, being either laid on the traditional mud and grass roof over timber rafters or used to form the internal leaves of cavity walls. Overall the architecture points towards a contemporary sustainable vernacular for the region.

Environmental design

School buildings

School buildings consist of a series of classrooms and staff offices grouped in two parallel buildings planned around an open courtyard, which provides play areas and additional secure outdoor teaching spaces.

The buildings, appositely separated to avoid overshadowing, take maximum advantage of the unique solar potential of the high altitude location by using glazed south-facing facades to gather the sun's energy, and high thermal inertia walls to store the gained heat.

On winter mornings the daytime teaching areas are quickly heated up by means of combining optimal 30° south-east orientation with fully glazed solar caption facades.

In summer, operable windows and roof lights allow cross-ventilation for cooling and fresh air.

All classrooms are entered from the courtyard via a lobby, which provides a thermal buffer. Each classroom has a quiet warm corner, with a small stove on a concrete floor that is only used on days of extreme cold weather. Timber floors elsewhere and white-painted mud rendered walls are provided for maximum teaching flexibility in clear, uncluttered spaces.

Residential buildings

Residential accommodation for boarding pupils is an important part of the overall plan for the school, catering for children from remote communities

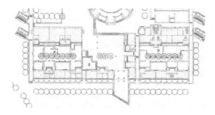

Figure 17.8 Plan of the school courtyards on the sides of the entrance and admin courtyard. All school buildings are arranged around a landscaped courtyard that serves as informal teaching space and breakout/play area

© Arup Associates

Figure 17.9 The school buildings are designed to gain maximum solar heat through the fully glazed south-facing facades and to provide high quality comfortable open spaces that can be flexibly used

© Caroline Sohie

Figure 17.10 Primary school: large clear spans have been created by using timber portal frames, which are earthquake resistant

© Caroline Sohie

Figure 17.11 In the junior school, a knee-brace solution has been introduced in the structural frame to reduce the amount of timber used

© Caroline Sohie

that would otherwise have very limited access to education. Similarly to the school buildings, the arrangement of the residential blocks is planned around a landscaped courtyard, which offers play areas and the possibility to grow their own food and vegetables to the little residents as part of their education programme. Each of the two buildings has shower rooms and a lounge. Sited as so to step towards south, the zoning of rooms within each of the blocks has been organized to provide good views out and cross-ventilation for each room, while at the same time maintaining privacy.

The residential buildings for boarding pupils are predominantly inhabited at night.

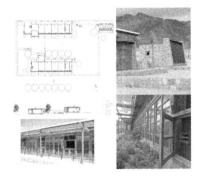

Figure 17.13 Internal view of a residential building. The back wall shows the ventilation dampers of the Trombe wall system in open position during winter

© Arup Associates

Figure 17.12 The residences are organized around a landscaped courtyard and face due south. Trombe walls provide passive solar heating to the spaces throughout the day, maintaining comfortable internal conditions

© Arup Associates

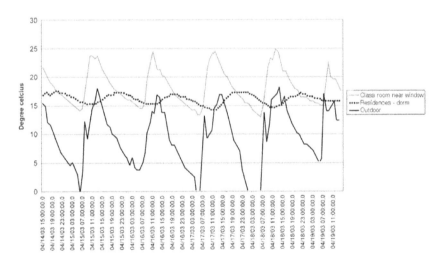

Figure 17.14 Temperature monitoring: graph showing the performance of two different passive solar heating solutions in the school buildings and in the residences, when late winter temperatures drop to below zero at night-time

© Arup Associates

They make use of Trombe walls for passive solar heating and are oriented due south. The thick walls that form the Trombe system are coated externally with dark heat-absorbing material and are faced with a double layer of glass, separated by 150mm to create a small air-space. Heat from sunlight passing through the glass is absorbed by the dark surface, stored in the wall, and conducted slowly inward through the masonry. Adjustable openings on the top and bottom of the thermal storage wall allow heat transfer from the heated air cavity to the room inside. This increases the efficiency of the system and ensures that the rooms are constantly kept at comfort conditions for the young occupants. Thick stone walls on the three sides of the block act as thermal storage for the heat gained through the Trombe wall, keeping the internal conditions almost stable, even when outdoor temperatures drop to below zero.

Monitoring of the thermal performance of the Trombe wall solution has revealed that indoor comfortable temperatures can be maintained during winter, without the need for additional heating.

In summer the Trombe walls are shaded and ventilated to prevent overheating, while the operable windows allow cross-ventilation, cooling and fresh air.

In all buildings occupants can control glare with internal light curtains and can use heavy internal light curtains to reduce unwanted heat losses from windows during night-time.

Structural design

Durability, flexibility and earthquake soundness are central aspects that govern structural design in this highly seismic zone. Ladakh is classified in seismic zone IV, the second highest category of the Indian Building Code. Although there have been no major earthquakes in the area in recent times, Ladakh has frequent tremors. The disasters caused by the Gujarat (2001) and Pakistan (2005) earthquakes showed the lack of well-engineered earthquake-resistant buildings in India and Pakistan and the devastation that can result, so that a strong seismic strategy was developed for the Druk White Lotus School.

The building structures use timber frames to resist the seismic loads and ensure life safety in the event of an earthquake. Traditional Ladakhi buildings are not engineered for seismic design, but with the application of some simple structural principles and details a huge improvement in earthquake safety can be achieved. One of the aims of the project is to act as an educational tool in the appropriate application of seismic design to traditional construction techniques, thus all structural solutions such as steel plates for the beam-column connection and cross bracings cables are exposed, revealing the simple yet effective solutions adopted.

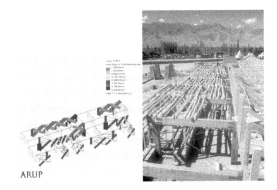

ARUP

Figure 17.15 The structural stability and earthquake design have been analysed with the GSA software. The solution adopted had to be easy to communicate to the local construction team and be built using hand tools and simple dry techniques

© Arup Associates

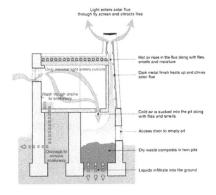

Figure 17.16 Ventilated Improved Pit (VIP) latrines provide a safe and hygienic improvement upon the traditional Ladakhi latrines, by the introduction of a solar flue that eliminates fly problems and odours

© Caroline Sohie

Figure 17.17 Ventilated Improved Pit (VIP) latrines: working principle of the solar flue and composting pits

© Arup Associates

All buildings have cavity walls on three sides. Granite blocks set in mud mortar are used for the outer leaf, while traditional mud-brick masonry is used for the inner leaf. This gives increased thermal performance and durability in comparison to the local rendered mud-brick walls. The Ladakhi-style heavy mud and straw roof is used and is supported by a timber structure that is independent of the walls. Steel connections and cross bracings provide earthquake stability.

Despite the complexity of the structural analysis, the design has been translated into simple solutions that have been easily understood by the local craftsmen and constructed within the constraints given by local materials and techniques.

Water and sanitation

Water is a scarce resource in Ladakh. Water for irrigation comes from the snowmelt, which is then used in a complex network of shared open channels in the intensely cultivated fields along the River Indus. To perturb the fragile water balance, the planned hydro-electric energy plant on the River Indus might impose a ban on water abstraction from the river, with the result that more households will revert to groundwater abstraction. This will have reverse effects in time, by reducing the capacity of the aquifer and having direct effects on the capacity of groundwater supply of the school too.

In fact the water cycle of the site relies on a solar-powered pump that delivers potable groundwater by gravity feed to a site that would otherwise be desert. When the pump is not in operation the solar panels charge batteries. The electricity generated can be used for lighting or small power supply to the school buildings.

The anticipated water demand of 60 litres per day per resident and 10 litres per day per student is comparably high by local standards, but is seen as a key aspect of the hygiene promotion programme that forms an important part of the education.

Key to the public health strategy for the school, traditional dry latrines have been adopted and enhanced to VIP (Ventilated Improved Pit) latrines. These eliminate fly and odour problems but do not require water. A double chamber

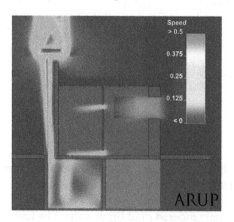

Figure 17.18 Ventilated Improved Pit (VIP) latrines: CFD analysis of ventilation performance of the solar flue

© Arup Associates

system with an integrated solar-driven flue allows their operation as com-posting toilets and produces humus that can be used as fertilizer. The solar flue is constructed as a chimney with a large south-facing metal sheet painted in a dark colour. The intense solar radiation induces the air cavity behind the metal sheet to warm up and subsequently rise upwards for buoyancy. This induces a continuous air circulation from the toilet into the pit and through the solar flue, which keeps the toilet well ventilated and healthier.

Wastewater from domestic uses is infiltrated via underground slotted pipes and a traditional open channel system along tree lines that shade and green the otherwise bleak high mountain desert environment. A second solar-powered pump uses a gravity supply to provide additional irrigation water to the whole site and to the vegetable gardens. These are aimed at varying the diet and teaching agricultural skills to the pupils.

Waste

In Ladakh everything is typically re-used. The population has for centuries survived by using Ladakh's natural resources carefully, without abusing them. For example, despite the long frosty winters, the scarce trees – apricot, willow and poplar – are not used for fuel. Rather they are carefully tended and their wood used only for construction or for tools. Dried animal dung is used for fuel and human waste as fertilizer: every house has a composting toilet and all 'waste' is recycled.

In the recent years, though, Ladakh has experienced the importing of packaged goods from India, such as plastic bottles and food wrapping. This has generated the escalating issue of how to dispose of this waste, given that there are no recycling facilities in the region. The result is that most plastic is currently buried or burnt, giving rise to a new phenomenon for the Ladakhis: pollution! A ban on plastic bags exists but it is not sufficient to alleviate the worsening waste situation.

Within this context the school seeks to reduce all waste produced, sepa-rating all waste streams, re-using waste materials where possible and only then send them to a local landfill. Plastic residue is buried in a nearby location that can be excavated in future, when a recycling facility is available in Ladakh. All organic waste is composted and wastewater from the food preparation in the kitchen feeds a reed bed system. Glass is re-used and paper waste is mixed in the composting.

Electrical power

Electricity in Ladakh is scarce and extremely unreliable. In the capital, Leh, a large diesel generator provides the homes with electricity at night, when the grid becomes unavailable. But in the villages, and at the school, power can be absent even for weeks.

A large hydro-electric plant is in construction in the area, by diverting the River Indus, but the works, planned to be completed years ago, have not advanced as thought and there is no certain date for the completion of the project.

This has led the team to come up with an alternative for the site: a phased photovoltaic installation that, when complete, will create a stand-alone local electricity grid to serve the electrical load for the full site.

The system is modular, with PV cells and inverter systems being installed within each building and connected into the site electrical supply. A modular battery system is being included, which will provide power in the hours of darkness. Future buildings will be fitted out with PV systems and integrated into the site grid as the school grows and funds become available. Additional battery capacity will also be added in the future, to support increasing site loads as the school develops.

The first stage of the installation will take place in summer 2008 and will include 54 monocrystaline solar panels, each rated at 170W peak power. These will be spread across three buildings, each with its own local inverter.

The full scheme will include 270 solar panels and an expanded battery system to provide full autonomy to the school.

The first phase of the PV installation has been donated by Arup Associates as a way to directly offset the carbon emissions generated by the business operation of the firm, by providing green power to the school.

At Arup Associates we have concerns over third-party offset, because of the lack of control and transparency over the carbon credit projects and their effective additionality. Instead, with this direct offset Arup Associates can monitor and have full control on the carbon reductions achieved on the site, confidently become carbon neutral and can link the practice even more closer with the local Ladakhi community in a long-term engagement of exemplary environmental awareness.

Next steps: the library

Whilst the normal construction of school buildings, labs and residences continues, funds are being raised for an exceptional building to be built at the core of the *mandala*: the library and assembly courtyard.

Eight exhibition and archive rooms face a circular central courtyard – modelled on the Buddhist symbol of the dharma wheel, which represents the 'unity of all things' – while a balcony offers access to the library.

The traditional mud roof helps with insulation and acoustics, while timber panelling and glazing provide a contrast with the granite walls that encircle the courtyard. High quality internal spaces and uniform daylight will be provided by the careful design of a fully glazed north-facing façade while the south-facing half of the circular building will become an open shaded balcony

Figure 17.19 Photovoltaic panels power the water supply system and provide the site with electricity for lights and computers

© Arup Associates

Figure 17.20 Visualization of the new library building: when complete it will become the heart of the masterplan

© Arup Associates

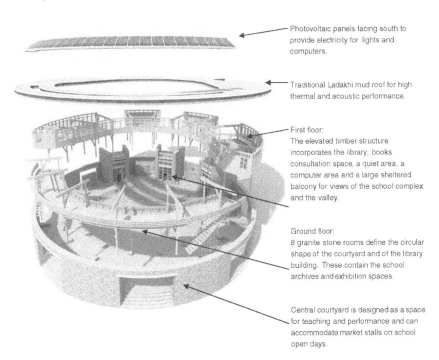

Photovoltaic panels facing south to provide electricity for lights and computers.

Traditional Ladakhi mud roof for high thermal and acoustic performance.

First floor:
The elevated timber structure incorporates the library, books consultation space, a quiet area, a computer area and a large sheltered balcony for views of the school complex and the valley.

Ground floor:
8 granite stone rooms define the circular shape of the courtyard and of the library building. These contain the school archives and exhibition spaces.

Central courtyard is designed as a space for teaching and performance and can accommodate market stalls on school open days.

Figure 17.21 Exploded view of the new library building under construction

© Arup Associates

for social interaction and informal learning. Photovoltaic panels installed on the rooftop will provide electrical power for lights and computers.

Conclusions

This project clearly demonstrates the value of progressive and sustained international collaboration and can be described as an example of how design can both symbolically and physically support a cause and maintain local traditions and culture, through a contemporary sensitive approach to design and resources.

The Arup approach has willingly avoided both the widespread internationalism of architecture and the praxis of mimicking local vernacular styles.

The design team has engaged in a long-term journey of partnership and exchange with the local community, both at technical and at cultural level, with an attitude open to learning from local experience and skills to lift both aspirations and expertise.

The typical attitude shown by 'Western designers' of imposing their designs and technological solutions, especially in developing countries and isolated communities, has always proven wrong and ineffective in time. Instead this project wants to demonstrate that intelligent buildings are those that learn from the skills and knowledge developed in centuries in a particular community; intelligent buildings should be the product of culture, climate and resources, reinterpreting these in a responsive and appropriate way, bringing innovation and improvement without depriving the design of its indigenous roots.

The Druk White Lotus School has become for the team a canvas for innovation and cultural understanding and has allowed Arup Associates to embark on a journey that goes well beyond the simple design and construction of a school campus, by supporting the vocational programme of the school and by directly offsetting the business carbon emissions of the practice with an important renewable energy investment.

Figure 17.22 Young students at the opening ceremony in 2001
© Caroline Sohie

Previously published in *Intelligent Buildings International*, 1(1): 82–96 (2009).

Chapter 18

Case study

Energy saving potential for a high-rise office building

Barış Bağcı

Within the context of this study, a walk-through energy audit was performed for a typical office building in Hong Kong. The focus of energy saving opportunities was mainly on mechanical ventilation and air conditioning (MVAC) and lighting. The emphasis was mainly on more energy-efficient equipment, a better control of the equipment, indoor air quality and human behaviour. Among others, the measures considered to reduce MVAC energy were replacing the air-cooled reciprocation chillers with water-cooled centrifugal chillers; installing variable speed drives (VSDs) for the secondary chilled water pump motors; maintaining the indoor air quality within reasonable limits while monitoring the indoor CO_2 concentration to adjust the fan speed of primary air-handling units (PAUs) by VSDs and re-adjusting the temperature setpoint to use less energy for air conditioning; and encouraging environmentally friendly human behaviour within the built environment, especially in hot and humid countries such as Hong Kong. The measures to reduce lighting energy were mainly replacing the lamps with energy-efficient lamps and preventing over-illumination. The results led to the conclusion that an annual saving of around 48% of MVAC and 62% of lighting energy is possible for this building. The measures discussed are also applicable to other similar office buildings in Hong Kong or similar areas, especially to old buildings. Considering that a tremendous amount of electrical energy is consumed by office and other commercial buildings, any measures to reduce the energy use in these sectors will contribute significantly to the reduction of CO_2 emissions and preservation of the environment.

Keywords: energy audit, energy saving, office building.

Introduction

The audited building is located in Tsim Sha Tsui district of Hong Kong. It is a 35-storey (including two basements) reinforced concrete building built in the early 1990s. It offers approximately 30,000m² of office space for some 2000 occupants. There are two levels of car park at the basement. From the

1st to the 33rd floor, it offers a first class office area for its tenants. Electricity is the only form of energy used by the service equipment in the building.

> Floor arrangement:
> Two basement levels: car park and air conditioning plant
> Ground floor to fourth floor: lobby area of the building
> 1st floor to 33rd floor: office area
> 34th floor and roof: water tanks, ventilation plants and goods lift.

The building operated 12 hours a day, six days per week. Daily operation of the air-conditioning and lighting system was controlled by building maintenance staff.

The performed study defines two main areas for energy saving opportunities: mechanical ventilation and air conditioning (MVAC) and lighting. The total monthly electricity consumption of the chillers was recorded by a sub-meter for one year's duration. The analysis considering the degree-days for each month shows a potential saving of 48% of MVAC energy. Furthermore, a saving of up to 62% of lighting energy is also possible.

In other terms, these measures mean an estimated saving of around 5371MWh/year for this building. The impact on the environment is a reduction of CO_2 emissions by 3670 tons/year – 0.7kg CO_2 per kWh (EPA, 2007). In financial terms, it means a saving of around US$627,400 per year (without considering the initial investment).

The next sections show how these savings can be realized.

MVAC

According to the Electrical and Mechanical Services Department in Hong Kong (EMSD) in 2004, air conditioning accounted for 30% of the total electricity consumption in Hong Kong. The electricity consumption by air conditioning had a growth of about 17% from 35,125 to 42,246 terajoules (TJ) from 1994 to 2004. The use of air conditioning is expected to grow further in view of the increasing population and economic activities. Therefore measures are needed to improve the energy efficiency, in particular, of air conditioning.

The analysis shows that in this area huge savings are possible. The energy conservation measures to improve the energy consumption of the building are explained under five subsections, which can be listed briefly as:

- Shutting the chillers down for four months, from November to February and using the fans only for fresh air supply during this period.
- Replacing the air-cooled reciprocating chillers with water-cooled centrifugal chillers.
- CO_2-based demand-controlled ventilation by installing variable speed drives (VSDs) in primary air-handling units (PAUs).
- Installing VSDs for the secondary chilled water pump motors.

Table 18.1 Monthly recorded chiller consumption and monthly degree-days

Month	Chiller consumption (MWh)	Monthly degree-days (ref. = 18.3°C)
January	180	13
February	185	17
March	360	150
April	360	200
May	500	290
June	540	330
July	550	370
August	600	420
September	570	360
October	430	260
November	270	90
December	230	31
Total	4775	2531

- Keeping the indoor air temperature at 25.5°C, as also recommended by the Hong Kong Government.

The total energy saving potential will be elaborated by going through those listed items one by one.

According to the recorded data in Table 18.1, the baseline equation can be found as y = 1040.5x + 178,449kWh by plotting the chiller consumption versus degree-days as shown in Figure 18.1.

This baseline equation will be used later in this section. Before that, it is necessary to expand Table 18.1 by adding more data, such as the number of chillers operated each month, total operating hours each month, total operating days each month and mean daily temperature. Except the last value, which is measured, the other data are obtained by calculation.

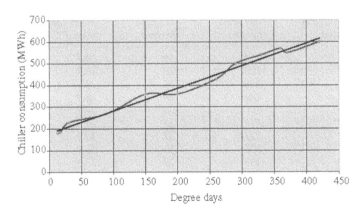

Figure 18.1 Baseline – chiller consumption vs. degree-days

The central air-conditioning system is equipped with four 300-ton (R22) air-cooled reciprocating chillers, each having a rated power input of 420kW, full load coefficient of performance (COP) of 2.5 and integrated part-load value (IPLV) of 1.41kW/ton. Operating one chiller consumes per month 420kW × 12h × (6/7) × 30 = 130MWh. Thus, the monthly chiller consumption also shows the average number of chillers operated per month.

The actual total operating hours per month can be found from (monthly chiller consumption)/[(number of chillers operated) × 420kW].

The actual total operating days per month is calculated by (total operating hours)/12. Note: this number is found to be larger than (6/7) × (days in a month) for a few summer months, which can be explained by the fact that during summer the chillers were sometimes operated more than 12 hours a day.

After performing those calculations, the obtained data are shown in Table 18.2.

Shutting chillers down for four months

After looking at Table 18.2, it is obvious that during the four months from November to February, the chillers can be shut down and filtered outdoor fresh air can be used instead of air conditioning. Since in Hong Kong these months have a low relative humidity, it is also not necessary to dehumidify the incoming air.

After taking this measure, the total monthly degree-days become 2531 − (13 + 17 + 90 + 31) = 2380.

Changing chiller type

Replacing the air-cooled reciprocating chillers with water-cooled centrifugal chillers is another measure which can be taken to reduce the electricity consumption.

Table 18.2 Monthly chiller operation data

Month	No. of chillers operated	Total hours operated	Total days operated	Mean daily temp. (°C)
Jan	2	214	18	18.7
Feb	2	220	18	18.9
Mar	3	286	24	23.1
Apr	3	286	24	25.0
May	4	298	25	27.7
Jun	4	321	27	29.3
Jul	4	327	27	30.2
Aug	4	357	30	31.8
Sep	4	339	28	30.3
Oct	4	256	21	27.0
Nov	3	214	18	21.3
Dec	2	274	23	19.3
Total	3.25 (avg.)	3392	283	25.2 (avg.)

Table 18.3 Efficiency recommendations for water-cooled chillers

Compressor type and capacity	Recommended IPLV (kW/ton)	Best available IPLV (kW/ton)
Centrifugal (150–299 tons)	0.52 or less	0.47
Centrifugal (300–2000 tons)	0.45 or less	0.38
Rotary screw ≥150 tons	0.49 or less	0.46

Table 18.3 shows efficiency recommendations for various types of chillers. The columns list the recommended level and the best available IPLV for that chiller type. Values are based on standard rating conditions specified in the Standard 550/590–98 of the American Refrigeration Institute (ARI).

Since the building location is included in the Pilot Scheme for Wider Use of Fresh Water in Evaporative Cooling Towers, launched by the government, it seems reasonable to switch to 300-ton (R134a) water-cooled centrifugal chillers, having a full load COP of 5.0 and IPLV of 0.45.

Figure 18.2 shows the area in Tsim Sha Tsui (TST), which is included in this scheme, where the building is also located (according to EMSD).

Assuming that the zero-load energy consumption of the chillers remains the same, the new baseline equation is calculated by simply changing the efficiency slope of the equation by the ratio of IPLV, which leads to:

$$y = (0.45/1.41) \times 1040.5x + 178449 kWh$$
$$\rightarrow y = 332.1x + 178,449 kWh$$

For total degree-days of 2380, the monthly average becomes 198.33 leading to a chiller consumption of $12 \times (332.1 \times 198.33 + 178,449) \cong 2931.8 MWh$ per year.

A poorly maintained freshwater cooling tower may become a breeding ground for Legionnaire bacteria. To ensure public health, the Code of

Figure 18.2 Part of TST, which is included in the pilot scheme (Google Earth, http://earth.google.com)

Practice for Prevention of Legionnaires' Disease should be followed and regular microbiological and water quality testing including Legionella and total bacteria count should be carried out (EMSD, 2003).

Introducing demand-controlled ventilation

Fan-coil units with primary treated fresh air were used for air conditioning the office areas. For each office floor, a constant supply of primary treated fresh air was delivered at a rate of 1500L/s by two PAUs on the respective floor. No major air quality complaints have ever been received and the CO_2 level is generally maintained at around 680ppm.

According to the steady-state mass balance equation, provided by ASHRAE Standard 62, Appendix C:

$$C_s - C_o = N/V_0 \tag{1}$$

where, C_s is CO_2 concentration in the space, ppm; C_o is CO_2 concentration in the outdoor air, ppm, which can be taken as 450ppm nominal value in Hong Kong; N is CO_2 generation rate, cfm (cubic feet per minute)/person; V_0 is outdoor airflow rate, cfm/person.

The CO_2 generation rate can be found as:

$$(680\text{--}450) \times 10^{-6} = N/(2 \times 1500\text{L/s})$$
$$\rightarrow N = 0.69\text{L/s}$$

which leads to 0.6831L/min or 0.024cfm per person, for 33 floors of office area and assuming 2000/33 persons in each office floor. This number corresponds to a MET (metabolic equivalent task) of 2.7 for an area with a light activity level, as shown in Figure 18.3.

2.7 MET per person seems to be high for an office area, where values of around 1.2 MET are common. This deviation can be explained by outdoor CO_2 concentration, which was assumed as 450ppm but might be more than this value in real life, especially in Hong Kong.

Nevertheless, it will be a conservative calculation, as the airflow rate is adjusted assuming a 2.7 MET per person. In reality, the supply airflow rate is expected to be lower than assumed here, which will be controlled by VSDs according to the measured actual CO_2 concentration, leading to less energy consumption than shown here.

According to ASHRAE Standard 62–1989, Section 6.1.3 together with Addendum 62f-1999, which revises the section: 'Comfort criteria with respect to human bioeffluents are likely to be satisfied if the ventilation rate results in indoor CO_2 concentrations less than 700ppm above the outdoor air concentration.'

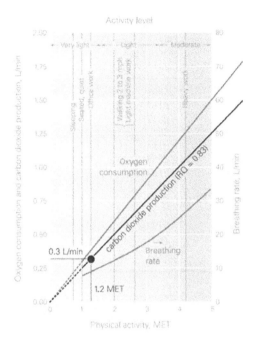

Figure 18.3 ANSI/ASHRAE Standard 62–2001, Figure C-2 (Murphy, 2002)

This obviously shows that the office areas in this building are over-ventilated and a reduction of the supply airflow rate would contribute to energy conservation.

For an outdoor CO_2 concentration of 450ppm, this would give an indoor concentration of 1150ppm, which is still acceptable. Keeping the indoor concentration at 1000ppm would still lead to a good indoor air quality (IAQ), which would make around 80% of the occupants feel comfortable.

For an excellent IAQ, we recommend keeping the CO_2 level at 800ppm, which will still significantly contribute to energy conservation.

From Equation (1), for an indoor CO_2 level of 800ppm, the supply airflow rate for one PAU can be calculated as:

$$(800\text{--}450) \times 10^{-6} = (0.69\text{L/s})/(2 \times V_0)$$
$$\rightarrow V_0 \cong 985\text{L/s}$$

The building has a total of 66 PAUs operating at a constant flow rate of 1500L/s. Each has a motor input power of 1.5kW leading to a power consumption of 66 × (total operation hours) × 1.5kW.

To calculate the energy consumption with the new IAQ setpoint, the fan speed is multiplied by a factor of 985/1500. The PAU power consumption is

Table 18.4 PAU consumption with new IAQ setpoint

Month	Total operation hours	PAU consumption (kWh)	PAU consumption (new) (kWh)
Jan	214	21,186	5999
Feb	220	21,780	6167
Mar	286	28,314	8017
Apr	286	28,314	8017
May	298	29,502	8354
Jun	321	31,779	8999
Jul	327	32,373	9167
Aug	357	35,343	10,008
Sep	339	33,561	9503
Oct	256	25,344	7176
Nov	214	21,186	5999
Dec	274	27,126	7681
Total	3392	335,808	95,087

multiplied by the cube of this factor since the fan power consumption P_f is proportional to the cube of the fan speed, v.

$$P_f \propto v^3 \tag{2}$$

The results are shown in Table 18.4.

Installing VSDs for secondary chilled water pump motors

The chilled water distribution system comprised constant speed pumps with differential bypass on primary circulation loop. The configuration of the building distribution loop was arranged as a direct return system with two-way valves at the terminal equipment.

There were a total of five constant speed secondary pumps running at a flow rate of 55L/s. The rated power of one pump motor was 40HP with 90% efficiency at full load and 1500rpm. This makes a power input of 40HP/0.9 $\cong$ 33.1kW for one motor.

The consumed energy at constant speed can be calculated from (number of pumps operating) $\times$ 33.1kW $\times$ (total operation hours).

It is possible to install VSDs for the five secondary chilled water (CHW) pump motors. To calculate the new energy consumption with VSDs, the following steps were taken.

It was assumed that the peak load, which occurred in August, will also be the full load operation point of the pumps, when the control valves in the secondary circuit were fully open. Proportional to temperature difference, the required CHW flow rate with VSDs in other months was estimated accordingly. Finally, the pump power consumption was multiplied by the cube of this factor, since the pump motor power consumption P_m is proportional to the cube of the flow rate, Q.

$$P_m \propto Q^3 \tag{3}$$

After these calculations, the obtained data are listed in Table 18.5. Summing up the consumption for all MVAC subsystems, the result becomes:
Current MVAC energy consumption:

4775MWh (chiller) + 335.8MWh (PAU) + 561.4MWh (pumps) = 5672.2MWh

MVAC energy consumption after applying saving methods:

2931.8MWh (chiller) + 95.1MWh (PAU) + 237.3MWh (pumps) = 3264.2MWh

Keeping the indoor temperature at 25.5°C

More energy can be saved by keeping the indoor temperature at 25.5°C instead of around 22°C, which was usually the case. Usually, 1°C increase of setpoint will lead to a conservation of around 3% in air conditioning energy. Thus, increasing the indoor air temperature from 22°C to 25.5°C will lead to a MVAC energy conservation of 10.5%.

When calculating this, the PAU energy consumption for the cooler for four months (total 25,846kWh) should not be included, since the operation purpose is only ventilation and not air conditioning, so the chillers will be shut down during this period. Thus, the additional energy saving would be:

0.105 × (3264.2 − 25.8)MWh ≅ 340MWh

Finally, after taking all described energy conservation measures, the MVAC energy consumption becomes:

3264.2 − 340 = 2924.2MWh

Compared with the current consumption of 5672.2MWh, this makes an energy saving potential of around 48%.

Lighting

44,000 600mm T12 fluorescent lamps with electromagnetic ballasts (EMB) (25W including gear loss) were mainly used for the office space. 500 60W incandescent light bulbs were installed in some circulation areas.

Table 18.5 Pump motor power consumption with VSDs

Month	Total operation hours	Mean daily temp. (°C)	Total CHW flow rate (L/s)	Energy consumption with const. speed (kWh)	No. of VSD pumps to operate	Required flow rate of each pump (L/s)	Pump motor speed (rpm)	Flow rate factor	Energy consumption with VSD (kWh)
Jan	214	18.7	8.15	35,417	–	–	–	–	–
Feb	220	18.9	12.22	36,410	–	–	–	–	–
Mar	286	23.1	97.78	47,333	2	48.89	1333	0.89	13,347.3
Apr	286	25	136.48	47,333	3	45.49	1241	0.83	16,238.6
May	298	27.7	191.48	49,319	4	47.87	1306	0.87	25,981.4
Jun	321	29.3	224.07	53,125.5	5	44.81	1222	0.81	28,233.1
Jul	327	30.2	242.41	54,118.5	5	48.48	1322	0.88	36,880.2
Aug	357	31.8	275	59,083.5	5	55	1500	1	59,083.5
Sep	339	30.3	244.44	56,104.5	5	48.89	1333	0.89	39,551.9
Oct	256	27	177.22	42,368	4	44.31	1208	0.81	18,012.9
Nov	214	21.3	61.11	35,417	–	–	–	–	–
Dec	274	19.3	20.37	45,347	–	–	–	–	–
Total	283 (avg.)	25.2 (avg.)	140.89 (avg.)	561,376	–	–	–	–	237,328.9

The lighting system of this building has energy saving potential by:

*   Replacing the T12 tubes with 549mm triphosphor T5 fluorescent tubes with quasi-electronic ballast (QEB) (14W) and removing the excess lamps to avoid over-illumination.
*   Replacing the incandescent light bulbs with 15W (18W with control gear loss) compact fluorescent lamps.

The next sections elaborate the potential saving in this area in more detail.

Replacing T12 tubes with T5 tubes and removing excess lamps

The measured illuminance for the office areas was 550–600lux, which is consistent with the Chartered Institution of Building Services Engineers (CIBSE) code for lighting, which recommends a maintained illuminance of 500lux for general offices.

The same illuminance can be achieved by using more energy-efficient lamps. The T12 tubes can be replaced by T5 tubes, which have a luminous output of 1250 lumen.

The new illuminance can be calculated from:

$$E = \frac{F \times N \times LLF \times UF}{Area} \tag{4}$$

where, E is illuminance; F is luminous output per lamp; N is number of lamps; LLF is total light loss factor (usually taken as 0.8); UF is utilization factor which depends on the room index and luminary (it is taken here as 0.5, an average value for a normal office area and light fittings with general purpose reflectors).

According to Equation (4) and for an office area of 30,000m², this leads to an illuminance of 733lux. To avoid over-illumination, $[(733–500)/733] \times 44,000 \approx 14,000$ lamps can be removed. Since each luminary contains three tubes, this can be done by removing one tube from each luminary.

Thus, by removing one third of the tubes and replacing the remaining tubes with T5 tubes, the average illuminance becomes:

$$E = \frac{1250 \times 30,000 \times 0.8 \times 0.5}{30,000} = 500lux$$

Table 18.6 shows the annual lighting energy consumption with T12 and T5 tubes (for 12 hours per day and 6 days per week).

After retrofitting and removing excess lamps, a saving of 4118MWh – 1573MWh = 2545MWh can be achieved.

Table 18.6 Comparison of lighting energy consumption with T12 and T5 tubes

Year	With T12 EMB (MWh)	With T5 QEB (MWh)
Total	4118	1573

Table 18.7 Comparison of lighting energy consumption with incandescent light bulbs and compact fluorescent lamps

Year	With incandescent light bulbs (MWh)	With compact fluorescent lamps (MWh)
Total	112	34

Replacing incandescent light bulbs with compact fluorescent lamps

Further savings can be achieved by replacing the 500 incandescent light bulbs (720 lumen output) by compact fluorescent lamps (800 lumen output). A similar calculation to that in the previous section leads to the results, which are shown in Table 18.7.

The additional savings would be 112MWh – 34MWh = 78MWh.

As a result, the total savings from lighting would be 2545MWh + 78MWh = 2623MWh. Compared with the current consumption of 4230MWh, the savings are about 62% of the current lighting energy consumption.

Additional savings can be achieved by installing occupancy sensors for common areas, which is not considered in this study.

The illustration in Figure 18.4 shows a comparison of electricity consumption by MVAC and lighting before and after the saving measures are considered.

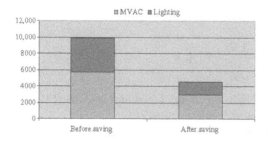

Figure 18.4 Annual energy consumption in MWhs

Conclusion

The measures introduced to reduce energy consumption by MVAC and lighting systems show that there is potential of saving more than half of the current electricity consumption for those areas. Table 18.8 summarizes the results.

Replacing the existing equipment could be considered because it is in use for more than 10 years already. But the investment costs are not considered in this study. However, the payback period is expected to be very short, considering the huge financial savings potential, so it is worth replacing the chillers and lamps; installing VSDs for the secondary CHW pumps; and installing CO_2 sensors and VSDs for the PAU fan speed control.

Furthermore, to achieve such savings does not depend only on equipment but also on people's behaviour. Especially in hot and humid countries such as Hong Kong, people need to get used to setting the indoor temperature at 25.5°C and make more use of natural ventilation during cool and dry days. They should be encouraged to wear light clothing inside the office, and not winter clothing while adjusting the air conditioner thermostat to 20–22°C.

Moreover, the companies should allow casual dress for their employees, such as the 'Cool Biz' campaign in Japan. This campaign, initiated by the Japanese Ministry of the Environment in summer 2005, aims to save energy by allowing casual dress and setting the indoor temperature to 28°C in summer.

So, the proposed indoor temperature of 25.5°C by the Hong Kong government should be comparatively easy for office workers to adopt. However,

Table 18.8 Energy saving potential summary (annual figures)

	MVAC energy	Lighting energy	Total
Current consumption (MWh)	5672.2	4230	9902.2
Consumption after measures are taken (MWh)	2924.2	1607	4531.2
Saved energy (MWh)	2748	2623	5371
In percentage	48%	62%	54%
In terms of saved money (US$)	321,000*	306,400*	627,400*
In terms of reduced CO_2 emissions (tons)	1924	1836	3760

*The amount was calculated by considering demand and energy charges according to China Light and Power Group bulk tariff in Hong Kong for on-peak periods and assuming a power factor of 0.9 (average demand is used instead of peak demand for easier calculation and the rebate for off-peak periods is not considered). The real amount might be slightly different. Monthly demand charge: 63.5HK$ per each kVA over the first 650kVA. Monthly energy charge: 68.6HK¢ per each unit over the first 200,000 units. (1HK$ = 7.8US$)

in most office areas in Hong Kong, the indoor temperature in summer is still around 20°C or even lower. This shows that, besides installing more energy-efficient equipment and implementing better control systems, human behaviour should be changed as well.

Previously published in *Intelligent Buildings International*, 1(2): 156–163 (2009)

Chapter 19

Case study

An intelligent building index awarded building

Lau Po-chi and Lam Yuen-man Bonnie

In the modern and competitive world, sustainable development and management should include engineering, managerial, financial and intelligent features and the ability to respond to the rapidly changing pace of technology and meet the increasing demand for a safe, healthy and comfortable home with quality. There are different kinds of tools in the industry to help property owners and operators access and improve their buildings. Among these assessment means, the intelligent building index (IBI) introduced by the Asian Institute of Intelligent Buildings (AIIB) emphasizes the holistic approach towards intelligent building assessment. This article aims to provide a brief review on the updates of the newly published IBI Version 4.0 and provides a case study on the assessment of the Intelligent Building Award, year 2009. It was the World Finance Centre Beijing (WFC), Beijing, PRC, which was granted the award under the IBI Version 3.0.

Keywords: elements; intelligent building index; modules; weighting.

Introduction

Notwithstanding the fast growth of intelligent building (IB) developments in the world today, there is still no fixed set of characteristics to define an IB in the market (AIIB, 2001). In order to give IBs in Asia a clearer and wider recognized definition, a group of professionals (including academics, engineers, architects, facilities managers and other related specialists) established the Asian Institute of Intelligent Buildings (AIIB) in December 2000. In the following year, Intelligent Building Index Version 2.0 was officially initiated (October 2001) (AIIB, 2005). AIIB believes that an IB is designed and constructed based on an appropriate selection of quality environment modules to meet the user's requirements by matching the appropriate building facilities to achieve long-term building value. It stated that each type of building (e.g. residential, industrial, commercial, hotels, schools, libraries, churches) should have its unique set of designed criteria to become an IB. Under this principle, AIIB allocates priorities in different modules for each type of building in the intelligent building index (IBI). It also believes that IB assessments should be done according to this rule. In this article, we briefly review the updates in IBI Version 4.0 and examine the assessment of the Intelligent Building Award, year 2009, World Finance Centre Beijing (WFC), Beijing, PRC, which was granted the award under IBI Version 3.0.

Development of IBI

In view of AIIB's IBI development, the Draft Intelligent Building Index Manual 1.0 was introduced to the market in May 2001. Its calculation was based on the Cobb–Douglas function, where nine sets of calculation modules were introduced in line with this standard. The first official IBI Manual Version 2.0 (IBIv2) was launched in October 2001 after feedbacks from the industry on IBI Manual 1.0 were gathered. In January 2005, the original nine-module index was revised to an index of ten calculation modules and the total assessment score was simplified to 100 maximum in IBI Version 3.0 (IBIv3). In January 2010, IBI was further updated and revised to IBI Version 4.0 (IBIv4).

Starting from IBIv2, many residential and commercial buildings have been assessed under the standards of this index.

IBI version 3.0

In this article, we have chosen an award-winning IB – WFC, PRC, which was granted the prize under IBIv3 in the year 2009 as a case study. In IBIv3, there are ten modules for the assessment of the building, and weighting varies for each index according to the type of building that was designed. The ten modules in IBIv3 are

1 environmental friendliness – green building,
2 space utilizations and flexibility,
3 home comfort,
4 working efficiency,
5 culture,
6 image,
7 earthquakes, disaster and structure damages,
8 construction process and structure,
9 life cycle cost – operations and maintenance with emphasis on effective-
    ness and
10 health and sanitations.

As AIIB believes that each type of building has a unique set of designed criteria to become an IB, it had assigned priorities (weights) in different modules for each type of building. The weights for different types of buildings in IBIv3 are shown in Table 19.1.

Table 19.1 Weights of different types of buildings in IBI Version 3.0

Type of building	M1	M2	M3	M4	M5	M6	M7	M8	M9	M10
Commercial	7	8.5	7.5	9	6	8.5	6.5	7	6	8
Hospitals	7	5	7	6	2	4	8	7.5	1	9
Residential	6.5	2	9	4	7	2	7	8	3	9
Hotels	4	6	9	5	4	7	6.5	8	3	8
Educational institutes	7	8.5	6.5	9	8	5	6.5	6	4	8

Table 19.2 Building ranking in IBIv3

IBI score	Ranking	Description
80–100	A	Distinction building
60–79.9	B	Credit building
50–59.9	C	Satisfactory building
35–49.9	D	Fair building
1–34.9	E	To be improved

According to the above assessment standard, the ranking of rated buildings is classified into the different levels shown in Table 19.2.

WFC – Intelligent building award, year 2009

WFC was awarded the Intelligent Building Award, year 2009. It is the first building in PRC to receive such an honour. In this assessment, WFC was rated grade A (distinction building) according to the standards of AIIB's IBIv3. The building was designed by the internationally renowned master architect Cesar Pelli, i.e. the same architect of the World Financial Center in New York, the Petronas Twin Towers in Kuala Lumpur and the Two International Finance Centre in Hong Kong (another Intelligent Building Award, year 2005 by AIIB).

Location and size

WFC is located in Beijing Chao Yang Central Business District. The whole development was completed in 2009. It is about 24 km from Beijing Capital International Airport. It is a twin block office tower with 24 storeys each at 100 m height situated on a site of 21,483 m² (231,243 ft²). The development has an approximate total floor size of 252,000 m² (2,712,858 ft²). The standard GFA for the typical office floor is about 4,400 m² (47,360 ft²) and 1,278 car parking slots are provided inside the development.

Figure 19.1 WFC located at CBD Beijing

The assessment

AIIB was invited by the developer to conduct an IBI assessment on the whole development in 2007. A team of three professional members carried out the first-stage pre-audit assessment visit in early April 2008. At this stage, the structure of the two buildings was basically completed, but most of the electrical, mechanical and building service equipments were still under construction. The audit team, based on IBIv3, presented a Stage I report and comments to the developer in August 2008. Although the development was not yet completed at this stage, a guide score of 88 was already obtained from the audit team.

In August 2009, the AIIB audit team organized the second (official) site visit to the development when most of the fitting-out of the building was completed. WFC was granted a distinction rating in this assessment.

The results of this second or final assessment were tabulated in accordance with IBIv3, and the scores of each of the ten modules are given in Table 19.3.

According to the comment of the audit team, this development is very strong in terms of environmental friendliness (score of 94.3), work efficiency (score of 96.2) and safety (score of 94.2). Such characteristics render the whole development as qualifying for a Class A office building. In addition, the project was highly rated according to LEED™ of USGBC. All the other modules, except for M10, were rated over 90 in this assessment. M10 also returned a score of 87.8, which is very close to the score of 90.

Structural layout and design

As to the structural design of the building, it is on a layout of column-free plan. Each of the two towers has 24 floors with 17 typical office levels and

Table 19.3 Assessment results for WFC

Module	Description	Weight	Score
MI	Green index	7	94.3
M2	Space index	8.5	93.3
M3	Comfort index	7.5	90.8
M4	Work efficiency index	9	96.2
M5	Culture index	6	90.6
M6	High-tech index	8.5	90.3
M7	Safety and structure index	6.5	94.2
M8	Management practice and security index	7	91.6
M9	Cost-effectiveness index	6	N/A*
M10	Health and sanitation index	8	87.8
IBI	Final score		92.1

*M9 cost-effectiveness index is the rent-to-cost ratio that should be calculated by total cost expenses per year to total rent per year in this assessment. Due to commercial secrets, the assessment in this part is not applicable.

three trading levels. The floor loading is 400 kg/m^2 for the typical office floor, 500 kg/m^2 for the trading floor and 1,200 kg/m^2 for the special zone around the building core.

Vertical transportation

WFC has 17 high-speed passenger lifts in three zones (each tower). The waiting time for using vertical transportation is less than 30 s during peak hours, which is considered to be very efficient in Beijing today. There is one executive lift (each tower) and two service lifts serving all floors (each tower), while four other separate lifts serve the car park levels between the basement and 1/F.

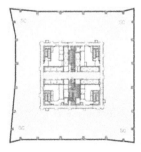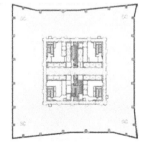

Figure 19.2 WFC is a twin tower office building with column-free design

Electricity and emergency backup

WFC has dual-power supply from two separate sources: dual-power supply risers in the building, and installed capacity 25,200 KVA including 10,000 KVA (4 × 2,500 KVA transformers) for office floors and trading floors. The load distribution is 100 VA/m² for typical office floors and 140 VA/m² for trading floors. Also, emergency power supply is provided.

Ventilation

Like many other new office buildings in the world today, WFC installed the VAV system with 24-h chilled water supply. For better human comfort and working efficiency, the building is designed with a fresh air supply rate at 40 m³/p/hour for typical office floors. According to the comment of the AIIB audit team, the car park and chiller plant room were also installed with high-efficiency chillers that used environmentally-friendly refrigerants.

Human comfort and the green index

One distinctive building feature in WFC is the 'winter garden' that lies between the two office towers. With this all-weather winter garden, the tenants of WFC are able to see green plantations even during heavy snow in winter in Beijing.

In Beijing, this development is very strong in terms of environmental friendliness, work efficiency and safety. Such characteristics render the whole development an environment very suitable for Class A offices and business development. At the same time, the project was highly rated according to LEED™ of USGBC. WFC meets LEED gold standards, which is a globally recognized assessment system for sustainable buildings formulated by the US Green Building Council. Also, WFC achieves the requirements of HK-BEAM, which is the building environment assessment method (BEAM) established

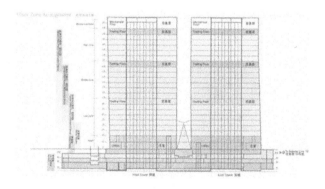

Figure 19.3 Winter Garden is built between two office towers

by the HK-BEAM Society. BEAM provides authoritative guidelines for the architectural and real-estate industries. It also leads the way in encouraging the adoption of groundbreaking measures that significantly reduce buildings' environmental impact while providing top quality workspaces for staff. Buildings that succeed in satisfying BEAM's stringent standards receive certification from the HK-BEAM Society. It is the landlord's aim to offer tenants new levels of safety, health, comfort, energy efficiency and sustainability in the buildings in China.

Reasons for being granted the award

WFC is a high-quality commercial building in Beijing. Besides its high-standard structural design, WFC also owns a comprehensive building management system with a fully digital security system with significant numbers of CCTV cameras. Although the building was still not launched into full operation at the time of assessment, its temporary building automation controls and security centre and the telecommunication plant room had already acted very professionally in the eyes of the audit team. The concept of a winter garden that provides tenants with an all-weather indoor green area is also rare in the nation. As a result, WFC has become the first building to receive the honour of the Intelligent Building Award of the Year in PRC.

Intelligent building index 4.0

IBI may claim to be the first quantitative and comprehensive assessment tool for IB rating in the world today. After the introduction of IBI in 2001, AIIB has been updating this index almost every four years for perfection (Version 2.0 in 2001, Version 3.0 in 2005 and Version 4.0 in 2010). The index emphasizes a holistic approach towards IB assessment. It introduces a multi-module assessment scheme weighted by the Cobb–Douglas utility function, which can simulate the non-linear judgement criterion of human beings (AIIB, 2005). Retaining this multi-module concept, IBIv4 has renamed some of the modules to reflect more accurately their concerns. In IBIv4, there are altogether ten modules and 377 numbers of elements. Also, a User Guide that allows users to pre-assess their building's intelligence is provided in this version.

In addition, IBIv4 provides the first electronic version of the IBI for assessment (a diskette with the worksheets is presented). This approach not only allows more user-friendly hands-on trials but also enables sensitivity analyses on the impact of a change of any individual element on the overall score. Users can try to pre-assess their buildings' intelligence and identify the SWOT (i.e. strengths, weaknesses, opportunities and threats) of their buildings in terms of building intelligence. With the help of this exercise, landlords can improve the intelligence of their buildings by the results obtained from the pre-assessment. Of course, certification can also be provided by AIIB on an approval process. In the assessment exercise, users can simply select the type of building and enter the scores on the individual elements. A combined score representing the overall score of the building is provided, an example of which is shown in Table 19.4. Users can also exclude any modules or elements by setting them as not applicable.

In the new version, the scoring system for all the elements is calibrated to meet a standardized scoring scale that helps benchmark the performance and interpret the context of the IBI facilities interpretations of the building. For a more accurate reflection of the index, IBIv4 has renamed the original Module 6 high-tech image index as the e-services and technology index. After the amendment, the ten new modules of IBIv4 are:

Module 1:   green index.
Module 2:   space index.
Module 3:   comfort index.
Module 4:   working efficiency index.
Module 5:   culture index.
Module 6:   e-services and technology index.
Module 7:   safety and structure index.
Module 8:   management practice and security index.
Module 9:   cost-effectiveness index.
Module 10:   health and sanitation index.

In this revised index, elements are grouped into three stages of development: (1) design, (2) construction and installation (including elements in stage (1) plus materials, specifications, workmanship), and (3) post-occupation stage (including elements in development stage (1) and (2) plus operations, maintenance, repairs and management). In other words, appropriate property management is crucial throughout the life of an IB (AIIB, 2010).

The combined score of IBIv4 can be simplified in Table 19.4.

Under IBIv4.0, a final IB index for a building can be obtained by aggregating in the Cobb–Douglas function. The scores of the ten modules are composed of 377 numbers of elements, which are shown in Table 19.5.

In this way, there are different combinations of modules in different priorities for different types of buildings. Once a module has been selected, a group of pre-selected facilities will be assigned accordingly. The choice of the designer towards a particular type of building only relies on the suggestions in Table 19.5. Thus, it would be possible to have one facility that belongs to more than one module. In other words, all the facilities are dependent on two factors: (1) the priority of the module in relation to the type of building, and (2) the availability of funding. IBI is in fact set up under this model.

Table 19.4 Combined score of the intelligence building index (an example)

Sub-index (module) code	Module weight (Y) (%)	Score (M)	CD weighted score (I)
GRI	9.46	10	1.2434
SPI	11.49	20	1.4107
COI	10.14	30	1.4116
WEI	12.16	40	1.5662
CUI	8.11	50	1.3733
ETI	11.49	60	1.6005
SSI	8.78	70	1.4523
MSI	9.46	80	1.5136
CEI	8.11	90	1.4403
HSI	10.81	90	1.6266
Total weight	100.00	Type: commercial building	
Total weighted score (I)			43.8940

Table 19.5 Ten modules of IBIv4

Module	Index name	No. of elements
M1	Green	80
M2	Space	18
M3	Comfort	52
M4	Working efficiency	72
M5	Culture	15
M6	e-Services and technology provisions	39
M7	Safety and structure	32
M8	Management practice and security	35
M9	Cost effectiveness	3
M10	Health and sanitation	31
	Total	377

Scoring standards

According to IBIv4, the IB score should range from 0 to 100. As '0' is mathematically not definable by the Cobb–Douglas function, the auditor can give a score of 0.000001, which would also significantly reflect the seriousness of the failure of that one particular item in the overall score.

The proposed scoring standards of IBIv4 divide the score into eight categories, from 5 to 100. The scoring standards help auditors and owners interpret the results of the assessment, and analyse the SWOTs of the buildings. IBIv4 believes that both the quality and extent of provision of the element should be assessed. When the assessed element achieves an excellent-quality level and an extent of substantial provisions, it would receive a score of 90. If further excellence is found, bonus scores can be given up to 100. If it is only a good-quality element that is not yet substantially good enough, a score of not higher than 70 should be given. A score equal to 60 is a yardstick for an element provision that just fulfils the latest statutory requirements, which are often the minimum. In case the provision is much below the latest statutory requirements, such as that in old buildings, the provisions may even result in illegality, hazards or very undesirable consequences, i.e. a score not higher than 5 (Table 19.6).

Sources of information

In IBIv4, there are several keys or indexing on each element for ease of reference: (1) sources of information, (2) disciplines, (3) stages of development, and (4) dimension of intelligence. These sources of information are grouped into six categories that help the auditors and owners find the required information to carry out the assessment. Such items are:

1   reports and drawings,
2   client,
3   visit on site,
4   test on site,
5   professional judgement and
6   log book.

IB ranking

According to the assessment, buildings can be ranked with the score. In IBIv4, the ranking arrangement has also been updated from that of IBIv3. Details of the updates are shown in Table 19.7.

Table 19.6 Scoring standards in IBIv4

Score	Scoring standards
0	Not allowed
5	Below statutory requirements and results in illegality or hazards or very undesirable consequences
10	Worst conditions or designs, etc., or undesirable consequences
40	Fair to bad conditions or designs, etc.
50	Lack of recommended items but not statutorily required; or just fulfilling old code
60	Just fulfilling the latest statutory requirements
70	Providing good but not substantial enough provisions
90	Excellent and substantial provisions
100	Excellent and substantial provisions plus bonus scores

Concluding remarks

This article has given a brief review on the updates of IBIv4. The case study that was an assessment under the standards of IBIv3 has actually shown the importance of appropriate support from the owners to the rating and the sustainability of IBs. Besides the difference in building intelligence ranking (Table 19.7), the main difference between IBIv3 and IBIv4 is the renaming of module no. 6 and the revision of element details in each module (see Table 19.8).

In this modern and competitive world, sustainable development and management should include engineering, managerial, financial and intelligent features and the ability to respond to the rapidly changing pace of technology and meet the increasing demand for a safe, healthy and comfortable home with quality, i.e. value for money services.

Table 19.7 Ranking of building intelligence by the combined score (I)

IBIv4 combined score (I)	IBIv3 combined score (I)	Ranking	Description
90–100	80–100	A	Distinction building
70–89.9	60–79.9	B	Credit building
60–69.9	50–59.9	C	Fair building
1–59.9	1–49.9	D	To be improved

Table 19.8 Module names and no. of elements difference in IBIv3 and IBIv4

Module	IBIv3	No. of elements	IBIv4	No. of elements
M1	Green index	75	Green	80
M2	Space index	18	Space	18
M3	Comfort index	50	Comfort	52
M4	Work efficiency index	80	Working efficiency	72
M5	Culture index	13	Culture	15
M6	High-tech index	38	e-Services and technology Provisions	39
M7	Safety and structure index	31	Safety and structure	32
M8	Management practice and security index	40	Management practice and security	35
M9	Cost-effectiveness index	1	Cost effectiveness	3
M10	Health and sanitation index	32	Health and sanitation	31
	Total no. of elements:	378		377

The explosion of technology coupled with the emergence of intelligent concepts means that more 'intelligence' can be incorporated in buildings (both commercial and residential) today. Of course, the final decision of whether to adopt IB concepts and the extent of 'intelligence' to be incorporated would depend on the owners and users' attitudes that are highly affected by the buildings' life-cycle costs and real actual benefits.

Previously published in *Intelligent Buildings International*, 3(1): 55–62 (2011)

Chapter 20

Case study

Intelligent building management

Alan Johnstone

In an intelligent building, the integrated building management system (iBMS) works in synergy with the other building components, from business systems to natural ventilation and light, and from the fabric of the building to the people who use it. The following case studies illustrate just how important the role of the iBMS is in intelligent buildings and why an integrated and forward thinking supply chain is essential, both pre and post occupancy.

Integrated systems

The application of building management systems in intelligent buildings requires seamless integration with other building systems. This is perfectly illustrated by the Trend system installed in the spectacular Radisson Blue Frankfurt Hotel.

With its unusual circular shape, the Radisson Blue 'Blue Heaven' Hotel in Frankfurt makes an eye catching addition to the city skyline. Opened in November 2005, it also impresses with the high standard of comfort that it offers its guests. The iBMS plays an important part in this, providing individual control of the air conditioning in every single guestroom.

The 20-storey Blue Heaven Hotel has 450 bedrooms, a conference suite, a ballroom and – 18 floors up – a swimming pool. Every bedroom is air conditioned and comfort is ensured by an open protocol LON BMS outstation.

Customer satisfaction is very important to the hotel so a wall-mounted panel is provided in every room to allow guests to adjust settings to their own individual requirements.

When a guest checks in, the iBMS automatically starts controlling room temperature to the occupied setpoint. On check-out it reverts to a setback level. This is possible owing to the interface between the iBMS and the hotel's Fidelio front office system, data from which is communicated to the room controllers via a Trend OPC server.

Also connected to this are an Ethernet based '963' supervisor (the system's main operator interface), plus the OPC server and the outstations that control the boilers, chillers and main air-handling plant to match the demands from

Figure 20.1 Radisson Blue Heaven Hotel

the individual room controllers. The whole installation functions as a single integrated system.

Through the '963' supervisor, the technical manager and his staff can not only adjust the system control settings and view data relating to the HVAC plant but are also able to monitor the position of its 610 fire and smoke dampers. The use of open protocols is an important part of the solution as the iBMS interfaces with the electrical systems using LON and connects to the fire damper controls via Modbus.

The diagram below is an example of how this network may be configured to integrate multiple open protocols.

Authorised personnel are able to access system settings and monitored data from almost anywhere in the hotel using just a laptop running Internet Explorer to browse the '963' web pages. They can do this from any of the guestrooms, all of which have wireless LAN access to the Internet.

The supply chain

Intelligent buildings not only rely on the integration of building systems but also the successful integration of the supply chain. It is only when the design team is fully integrated that the greatest benefit can be delivered.

This integrated approach was essential in the design of the iBMS on one high-profile building. When Andy Murray beat Stanislas Wawrinka in a five-set thriller to reach the quarter final of the 2009 Wimbledon Championship, this was the first match to be played entirely under the Centre Court's new

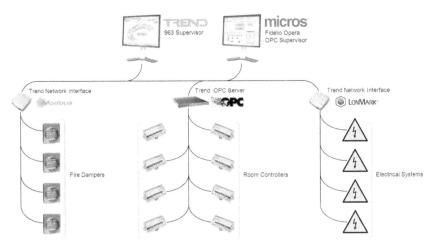

Figure 20.2 Integration network

retractable roof. In contrast to the highs and lows experienced by spectators, climatic conditions beneath the closed roof remained remarkably stable throughout the four-hour encounter.

Over the preceding three years, the All England Lawn Tennis Club (AELTC) and its main contractor Galliford Try Construction carried out a major upgrade of the Centre Court stadium. This famous sporting arena now has a larger capacity (15,000), more comfortable seats, new catering facilities, and a high-tech, translucent folding fabric roof. The new roof, and the sophisticated air management system that operates when it closes, allow play to continue during inclement weather.

During the Murray v. Wawrinka match the temperature and relative humidity (RH) levels within the stadium bowl hardly varied, ranging from 23.5 to 25.4°C and 49–52% RH (the outside temperature and humidity averaged 27°C and 58%). It took less than ten minutes to fully close the roof and only a further 20 for conditions beneath it to stabilise.

Crucially, the grass playing surface was kept completely free of moisture. Preventing condensation, and thereby ensuring that the court is non-slippery and thus safe to play on, is the overriding priority for the ventilation system and its Trend controls.

High up in the fixed section of the Centre Court roof are 14 air-handling units (AHUs), which are individually controlled and monitored by Ethernet-linked Trend outstations. The AHUs deliver a combined air flow of 120,000 l/s and discharge into three ductwork rings that circle the court. Tempered air is fed down into the stadium bowl through 64 elliptical nozzles and numerous diffusers, the nozzles being supplied via 18 variable air volume boxes. Further outlets push air across the underside of the roof to stop condensation forming on its surface and falling onto the court.

There are six temperature sensors embedded in the grass around the edge of the court and eight attached to the roof trusses. A further 40 combined temperature/humidity sensors are located round the stadium, many of them in the passages leading to the seating. As well as measuring conditions, they repeatedly calculate the dew-point temperature, their readings being averaged by the iBMS. The average dew-point is continually compared with the minimum grass and roof temperatures.

PACE Services installed the first Trend controls at Wimbledon some 25 years ago and has gone on to fit them throughout what is a large and complex site. Throughout the development and implementation of the innovative Centre Court roof project, PACE worked very closely with the AELTC, its main contractor Galliford Try Construction and its building services consultants M-E Engineers and Foreman Roberts. As a consequence it was able to make an important contribution to the design of the Centre Court controls and the project's overall success.

Natural resources

The integrated supply chain was also a component of the new control strategy at London's National Gallery which is now better equipped to present its old masters in the best possible light – quite literally. The site's Trend building management system has begun implementing a new strategy for controlling gallery roof blinds, one that will allow the collections to be viewed in natural lighting for longer periods, while protecting them from direct sunlight. As well as enhancing the viewer's appreciation of a painting, the use of daylight to provide illumination reduces the need for artificial lighting, thus saving energy.

The Trend iBMS provides close control and monitoring of environmental conditions throughout the National Gallery's main building. In doing so it plays a vital role in preserving the gallery's priceless collection of 13th–19th century Western European paintings, which include masterpieces from virtually all the great artists. Most of the building's 40 plus galleries have glazed roofs and here the system regulates the light levels through coordinated control of the picture lighting and the rooflight window blinds.

The iBMS will only switch on the artificial lighting if the optimum illumination level cannot be achieved by natural light alone, through control of the blinds. The gallery has two tracklight circuits, one comprising blue lights that simulate daylight and the other made up of conventional clear lamps. The first are brought on if the average light level falls to a preset value, and if this is not sufficient the second set is switched on as well.

When modulating the motorised blinds to achieve the desired light level, the controllers must ensure that no direct sunlight is admitted. They continually calculate the changing position of the sun and compute maximum and minimum safe opening limits for each set of blinds taking into account their orientation and slope, all to a resolution of less than one degree.

The blind control strategy was devised by building services consultant Andrew Reid & Partners who, like Trend, has worked with the National Gallery for many years. In a coordinated design approach, the program was tested using an architectural software package that includes solar simulation and supply settings. The system carries out extensive trigonometric calculations for around 100 sets of blinds of varying design and orientation and has provided significant energy savings.

Another crucial task performed by the Trend system is control of gallery temperature and humidity. Maintaining stable conditions is particularly important for picture preservation. Close collaboration between Trend and Andrew Reid & Partners has resulted in a number of strategy changes and control refinements through the life of the gallery to maintain environmental stability.

Building fabric

The intelligent systems in the National Gallery maximise natural light to provide an environment which is perfectly suited to the building's use and enhances the experience of the building's users while minimising its energy profile. Integrating intelligent building systems with innovative fabric design in this way can deliver extraordinary results.

The energy required to provide heating and cooling of the University of East Anglia's Elizabeth Fry Building, which features a TermoDeck fabric thermal storage/ventilation system, is so low as to be almost unbelievable –

Figure 20.3 National Gallery, London

particularly given that occupant satisfaction levels are remarkably high. The outstanding result is down to the iBMS and the time and effort that went into fine tuning based on performance data produced by the system itself.

The four-storey Elizabeth Fry building has a floor area of $3250\,m^2$ and contains offices, seminar rooms and lecture theatres. Though heavily used and mechanically ventilated it consumes just $41\,kWh/m^2/year$ for heating and cooling, which is well below even the official 'good practice' level for naturally ventilated premises. Indeed, there is possibly no other building in the UK that can match its energy efficiency.

When it was opened at the beginning of 1995, Elizabeth Fry was only the second building in the UK with a TermoDeck system. Developed in Sweden it comprises a series of interconnected hollow-core concrete slabs, through which air is delivered to the space. Heat from computers, lights, etc. radiates to the slabs and is recovered from the extract air. In hot weather, cool night air is blown through the slabs to reduce their temperature. The building has no refrigeration plant and relies entirely on this passive cooling effect.

The building's impressive levels of insulation, air-tightness and heat recovery have also been major factors in keeping down energy consumption, as has the installation of the Trend control system. The system controls and monitors the building's four air-handling plants and the small boilers which serve them. Data recorded by the iBMS, particularly historic logs of temperatures and other variables, have helped the university to gain a through-life understanding of the building's behaviour and the characteristics of the TermoDeck, allowing the implementation of a variety of energy-reducing control strategy changes.

The tuning of the system was carried out over a period of almost two years and resulted in both gas and electricity savings. Gas consumption more than halved, tumbling from $58\,kWh/m^2/year$ at the start of 1996 to its current level of $25\,kWh/m^2/year$ – equivalent to an annual gas bill of just £750. Tellingly, it is rarely necessary for the system to operate more than one of the building's domestic size boilers – and even this unit runs for less than 2,000 hours per year.

Post-occupancy evaluation surveys have shown that, despite the building's low energy usage, 70 per cent of its occupants are very satisfied with conditions. Even during severe heat waves, a comfortable environment has been maintained.

Engaging with building users

A desire to be more responsive to the occupants' needs and increase customer satisfaction has lead one UK high street bank to reconsider how its operational staff interact with building users. The result is a pilot of the latest in mobile user interfaces for their Trend Building Management System at Bishopsgate in London.

This groundbreaking project combines the latest in mobile tablet technology, secure 3G mobile networking and market leading building management systems to deliver real-time property performance data directly to building operatives.

The innovative solution uses the Trend 963 supervisor to web serve real-time building and plant performance data to rugged Windows 7 tablets anywhere in the building, allowing on-site engineers a mobile view of the building temperatures and settings that affect occupant comfort and energy consumption.

The BMS Tablet enables the building support staff to deliver a proactive service by bringing iBMS performance data to each desk. This reduces response times and significantly enhances the client experience for building users.

Maintenance and commissioning tasks are delivered more efficiently as engineers can effect plant changes from wherever they are in the building and monitor the results of the changes in real-time. This is particularly effective in large open-plan areas with restricted access and terminal units hidden above false ceilings, such as Bishopsgate.

The system is highly scaleable and portable, making iBMS data accessible from one building to another with no additional investment in infrastructure, indeed, operators can interact securely with their building wherever there is a mobile signal.

According to the bank's operations manager:

> We can now really interact with the building users, providing an at-desk hot and cold call service. By showing the occupant what is going on behind the scenes, the Property Services team are able to reduce calls as occupants become more aware of their environment and its constraints.

The BMS Tablet is a product of the IP capabilities of the iBMS, which is specifically designed to run on TCP/IP networks with embedded web capability to allow remote interrogation by web browsers, and recent developments in tablet technology.

The chosen device is the Motion CL900 which is a durable, lightweight and powerfully equipped tablet PC specifically designed for business applications. The CL900 is a touch screen Windows 7 tablet with 3G and wireless capability.

The Motion CL900 Tablet views the Trend 963 client using Microsoft Internet Explorer and connects to a dedicated host server using an SSL Agent. This means that the device can operate out of the box with no specialist software applications.

The 963 client provides the remote user with the full functionality of the main supervisor on a web page, allowing the user to view schematics, display graphs, adjust parameters, receive alarms and modify diaries in much the same way as the main operating station.

Figure 20.4 Tablet with BMS graphic

Implementing the BMS Tablet has some clear and considerable benefits for the building operator. The delivery of real-time data direct to operatives in the field, where and when they need it, maximises their productivity and significantly enhances both the time to deliver and effectiveness of fault responses. Maintenance provision is immediately more proactive and information can be presented clearly and concisely to building occupants, increasing user satisfaction.

The use of tablet PCs to make more effective use of iBMS data can allow building systems to perform seamlessly and thus free-up businesses to focus on operational issues and, in turn, have a significant impact on operational cost. The Royal Academy of Engineering (1998) estimates the cost ratio for a new building as 1:5:200 for construction, maintenance and operation. Any system which minimises maintenance and helps to reduce the operational element of the cost base is therefore extremely valuable.

There are also some significant productivity gains to be made; according to Microsoft (2005), 'organizations that have deployed tablet PCs have been able to significantly improve work processes and generate substantial returns on their investment in this technology'. Productivity gains are not limited to the tablet operator, BISRIA (2003) states that particular improvements from iBMS Maintenance accrue through improvements in occupant productivity, pointing to a number of studies that claim productivity gains of between 3 and 15 per cent through thermal satisfaction. Thermal satisfaction is a mix of how comfortable the building occupant feels and how effectively the building operator communicates and responds to issues.

The BMS Tablet helps to maintain comfort conditions at their optimum level by making building data clearly visible for all to see. It ensures that maintenance response is proactive, reducing downtime, and it allows the field operative to communicate effectively with the building user. This engagement with the user is the most important feature as it highlights how the user and building affect each other and improves the perception of comfort.

Designing buildings for people

One county council has gone a stage further by designing schools around the people who use them. In January 2010, Sutton Park Primary in Kidderminster and its 240 pupils moved into brand new school buildings designed by Worcestershire County Council's own Property Services team.

In contrast to the ageing premises they left behind, their current home is distinguished by a low-energy design and a strong emphasis on renewable resources, which include a wood chip-fuelled biomass boiler and a rainwater harvesting system. Good insulation and integration of the iBMS with the building systems has helped ensure a high level of energy efficiency. In fact, heating costs are less than half those of the old school, even though the new buildings have a larger floor area.

Teachers at the school are doing their best to ensure that their pupils do not take the world's natural resources for granted. They are being helped in this quest by the iBMS, part of whose function is to educate the children about the importance of energy conservation and sustainable living.

While the maintenance staff use the iBMS to ensure the best environment for learning, from temperature to CO_2 concentrations, within the schools, the pupils at Sutton Park can also look at iBMS collected data about their school through custom-designed, child-friendly graphic pages that Trend created at the council's request.

The iBMS has been incorporated into the school curriculum and interactive pages are displayed on whiteboards. Using simply worded text, animated graphics and a pictorial style familiar to a young audience, the children are shown how their classrooms are kept warm and their hand washing water is heated, and how rainwater is collected and used for flushing toilets.

The displays contain live and historic data collected by the iBMS, such as temperature, energy and rain water consumption. Another page, which aims to encourage energy saving behaviour, gives the electricity usage for each classroom area. In the school's library there is a Trend EnergyEYE, which

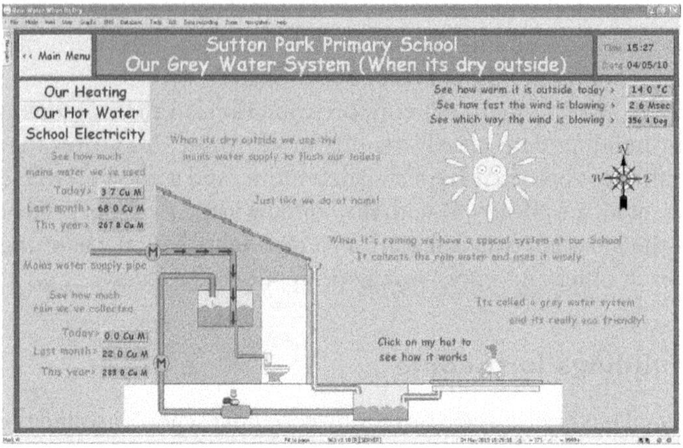

Figure 20.5 Child-friendly BMS graphics

provides a permanent display of the latest energy consumption figures, and their CO_2 equivalents, on a large plasma screen.

Following some very positive feedback, the council has adopted the child-friendly graphics used at Sutton Park for other schools and has established this innovative teaching aid as part of the curriculum.

Bibliography

Abdul-Wahab, S.A. (2011) Sick building syndrome in public buildings and workplaces. In D. Clements-Croome *The Interaction Between the Physical Environment and People*, Berlin: Springer Verlag, Chapter 13.

AIIB, 2001, The Intelligent Building Index Manual 2.0, Asian Institute of Intelligent Buildings, Hong Kong.

AIIB, 2005, The Intelligent Building Index Manual 3.0, Asian Institute of Intelligent Buildings, Hong Kong.

AIIB, 2010, The Intelligent Building Index Manual Version 4.0, Asian Institute of Intelligent Buildings, Hong Kong.

Akhlagi, F. (1996) Ensuring value for money in FM contract services, *Facilities*, 14(1/2) January/February: 26–33.

Alvarsson, J.J., Weins, S. and Nilsson, M.E. (2010) Stress recovery during exposure to nature sound and environmental noise, *International Journal of Environmental Research and Public Health*, 7(3): 1036–1046.

Anderson, J. and French, M. (2010) Sustainability as promoting well-being: psychological dimensions of thermal comfort. Personal communication. Institute of Well-Being at the University of Cambridge.

Aouad, G., Bakis, N., Amaratunga, D., Osbaldiston, S., Sun, M., Kishk, M., *et al.* (2001) An integrated life cycle costing database: a conceptual framework. *ARCOM Conference*, The University of Salford.

Arbib, M.A. (2012) Brains, machines and buildings: towards a neuromorphic architecture, *Intelligent Buildings International* 4(3): 147–168.

ARI Standard 550/590–1998, Standard for water chilling packages using the vapor compression cycle.

Arnold, D. (1999a) The evolution of modern office buildings and air-conditioning. *ASHRAE Journal*, June, pp. 40–54.

Arnold, D. (1999b) Air-conditioning in office buildings after World War II. *ASHRAE Journal*, July, pp. 33–41.

ARUP (2003) *Intelligent Buildings*. Arup. Available at: http://arup.com/communications/skill.cfm?pageid=4311. [Accessed 21 Feb. 2009]

ASHRAE (1995) HVAC Applications. *ASHRAE Handbook of the American Society of Heating, Refrigeration and Air-Conditioning Engineers*. Atlanta, GA: American Society of Heating, Refrigeration and Air-conditioning Engineers.

ASHRAE (1996) HVAC Systems and Equipment. *ASHRAE Handbook of the American Society of Heating, Refrigeration and Air-Conditioning Engineers*.

Atlanta, GA: American Society of Heating, Refrigeration and Air-conditioning Engineers.

ASHRAE (2004) *Standard 55–2004, Thermal environmental conditions for human occupancy*. Atlanta, GA: American Society of Heating, Refrigeration and Air-conditioning Engineers.

ASHRAE (2010) *Standard 55–2010, Thermal environmental conditions for human occupancy*. Atlanta, GA: American Society of Heating, Refrigeration and Air-conditioning Engineers.

ANSI/ASHRAE Standard 62–2001, Ventilation for acceptable indoor air quality.

Bahr, N.J. (1997) Fault tree analysis. Chapter 7 in *System Safety Engineering and Risk Assessment: a practical approach*, New York: Taylor & Francis.

Baker, N., Fanchiotti, A. and Steemers, K. (1993) *Daylighting in Architecture: a European reference book*. London: James & James.

Bako Biro, Zs., Clements-Croome, D.J., Kochhar, N., Awbi, H.B. and Williams, M.J. (2012) Ventilation rates in schools and pupils' performance, *Building and Environment*, 48, February, 215–223.

Bako Biro, Zs., Kochlar, N., Clements-Croome, D.J., Awbi, H.B. and Williams, M. (2007) 'Ventilation rates in schools and learning performance'. In: *Proceedings of CLIMA 2007 – WellBeing Indoors*, The 9th REHVA World Congress, Helsinki, Finland 1434–1440.

Bako Biro, Zs., Kochhar, N., Clements-Croome, D.J., Awbi, H.B. and Williams, M. (2008) Ventilation rates in schools and pupil's performance using computerised assessment tests. In: *Proceedings of the 11th International Conference on Indoor Air Quality and Climate*, August 17–22, 2008, Copenhagen, Denmark.

Baldry, C. (1999) Space: the final frontier, *Sociology*, 33(3): 1–29.

Baue, B. (2006) *Opening the Umbrella of Socially Responsible Investing to Include Energy Efficient Mortgages*. Brattleboro VT: SRI World Group, Inc. [Online] Available at: http://www.socialfunds.com/news/article.cgi/1934.html [accessed 8 October 2009].

Berglund, B. and Gunnarsson, A.G. (2000) Relationships between occupant personality and the sick building explored, *Indoor Air*, 10: 152–169.

Bernstein, H. and Russo, M. (2010) Personal Communication. McGraw-Hill Construction.

Best Practice Programme (2000) *Energy Consumption Guide 19: Energy Use in Offices*. Crown Offices, UK.

BISRIA (2003) *BG4/2003 BMS Maintenance Guide*.

Blanchard, B.S. (1992) *Logistics and Engineering Management*, fourth edition. New Jersey: Prentice Hall.

Blyth, A. and Gilbe, A. (2006) *Guide to Post Occupancy Evaluation*. HEFCE, AUDE and the University of Westminster.

Blyth, A. and Worthington, J. (2001) *Managing the Brief for Better Design*. London: E & FN Spon.

BMI (2003) *Review of Occupancy Costs 2003*. BMI Special Report Serial 322, Building Market Information. The Royal Institution of Chartered Surveyors.

Booy, D., Liu, K., Qiao, B. and Guy, C.A. (2008) *Semiotic Model for a Self Organising Multi-Agent System*. In *DEST2008 – International Conference on Digital Ecosystems and Technologies*. Phitsanulok, Thailand: IEEE.

Bowen, P. (2005) *Integrated Approach for Information Communication Technology (ICT) and Control System Infrastructures within Buildings: An*

Independent Study. Converged Building Technologies Group. Available at: http://www.intelligentbuildings.com/PDF/library/smartBuildings/CBTG_ROI_M odel.pdf [accessed 8 February 2009].

Boyden, S. (1971) Biological determinants of optimal health. In D.J.M. Vorster (Ed.) *The Human Biology of Environmental Change.* Proceedings of a conference held in Blantyre Malawi, 5–12 April. London: International Biology Programme.

Braungart, M. and McDonough, W. (2009) *Cradle to Cradle.* London: Vintage.

BRE, 2004, *Ecopoints: A Single Score Environmental Assessment,* BRE, Watford, UK.

BRE (2005a) *Report 476: Achieving Whole Life Value in Infrastructure in Buildings.* Watford: Building Research Establishment.

BRE (2005b) *BREEAM: Building Research Establishment Environmental Assessment Method.* Watford: Building Research Establishment.

BRE (2008) *BRE Environmental Assessment Method (BREEAM).* Watford: Building Research Establishment.

BRE (2012) *National Calculation Method.* Watford: Building Research Establishment. Available from: http://www.ncm.bre.co.uk/page.jsp?id=1 [accessed 3 September 2012].

British Council for Offices (2009) *Guide to Specification.* London: BCO.

BSI (1986) BS 5760:1986 *Reliability of Systems, Equipment and Components.* British Standards Institution, UK.

BSI (1991) BS 5925:1991 *Code of Practice for Ventilation Principles and Designing for Natural Ventilation.* British Standards Institution, UK.

BSI (1995) BS EN ISO 7730:1995 *Moderate Thermal Environments. Determination of PMV and PPD Indices and the Specification for Thermal Comfort.* British Standards Institution, UK.

BSI (2000) BS ISO 15686:2000 *Building and Constructed Assets: Service Life Planning,* parts 1–6. *(General Principles Part I – published).* British Standards Institution, UK.

BSI (2004) BS EN 13779:2004 *Ventilation for Non-Residential Buildings. Performance Requirements for Ventilation and Room-conditioning Systems.* British Standards Institution, UK.

Building Regulations (2000a) Part L2A: *Conservation of Fuel and Power in New Buildings other than Dwellings.* HMSO, UK.

Building Regulations (2000b) Part L2B: *Conservation of Fuel and Power in Existing Buildings other than Dwellings.* HMSO, UK.

Building Regulations (2010) *Building and Buildings, England and Wales,* Statutory Instrument No. 2214, HMSO, UK.

Burge, P.S., Hedge, A., Wilson, S., Bass, J.H. and Robertson, A. (1987) Sick Building Syndrome: a study of 4373 office workers, *Annals of Occupational Hygiene,* 31: 493–504.

Burklin, B. (2007) A collaborative model for the construction industry. In K. O'Donnell and W. Wagener (eds) *Connected Real Estate: Essays from Innovators in Real Estate, Design and Construction.* Cisco. Walmer: Torworth.

Burr, A. (2008) *CoStar Study finds Energy Star LEED Buildings Outperform Peers.* Bethesda, MD: CoStar Realty Information Inc. [Online] Available at: http://www.costar.com/News/Article.aspx?id=D968F1E0DCF73712B03A099E0 E99C679 [accessed 8 October 2009].

Burton, E.J., Bird, W., Maryon-Davis, A., Murphy, M., Stewart-Brown, S., Weare, K. and Wilson, P. (2011) *Thinking Ahead: Why We Need to Improve Children's Mental Health and Wellbeing*. Edited by Rachael Jolley. London: Faculty of Public Health.

Business Vantage/Construction Clients' Group (2009) *Equal Partners – Customer and Supplier Alignment in Private Sector Construction* [Online] Available at: www.businessvantage.co.uk/equalpartners/Equal Partners 2009.pdf

CABA Report (2007) *Introduction to Commercial Building Control Strategies and Techniques for Demand Response*. Canada: Continental Automated Buildings Association.

CABA Report (2008) *Bright Green Buildings: Convergence of Green and Intelligent Buildings*. Canada: Continental Automated Buildings Association.

Cabanac, M. (2006) Pleasure and joy, and their role in human life. In D. Clements-Croome (ed.) *Creating the Productive Workplace*, Oxford: Taylor& Francis, pp. 40–50.

Cabinet Office (2011) *Government Construction Strategy* paragraphs 2.28 and 2.44 and Action Plan items 6 and 11 [Online] http://www.cabinetoffice.gov.uk/sites/default/files/resources/Government-Construction-Strategy.pdf.

Cabinet Office (2012) *Newsletter* –see *Interim Report* and *Pilot Projects* [Online] https://update.cabinetoffice.gov.uk/sites/default/files/resources/construction%20newsletter%20Feb%202012.3.pdf.

Cao, M. and Wei, J. (2005) Stock market returns: A note on temperature anomaly, *Journal of Banking & Finance*, 29: 1559–1573.

Carbon Trust (2002) *Low Carbon Technology Assessment 2002: Making Our Investment Count*. London: Carbon Trust.

Carder, P. (1997) The interface manager's toolkit. *Facilities*, 15(3/4): 84–89, MCB UP Ltd.

Castleton, H.F., Stovin, V., Beck, S.B.M. and Davison, J.B. (2010) Green roofs: building energy savings and the potential for retrofit, *Energy and Buildings*, 42: 1582–1591.

Chappells, H. (2010) Comfort, well-being and the socio-technical dynamics of everyday life, *Intelligent Buildings International*, 2(4): 286–298.

Chrenko, F. (1974) *Bedford's Basic Principles of Ventilation and Heating*. Third Edition. London: H.K. Lewis and Co. Ltd, Chapter 8.

CIB (1999) *Agenda 21 on Sustainable Construction*. CIB report, publication 237.

CIBSE code for lighting. Available at: www.cibse.org/index.cfm?go=page.view&item=453.

CIBSE (1999a) *Technical Memorandum 24: Environmental Factors Affecting Office Worker Performance; a Review of the Evidence*. London: Chartered Institute of Building Services Engineers.

CIBSE (1999b) *Applications Manual AM12-small scale CHP in buildings*. London: Chartered Institute of Building Services Engineers.

CIBSE (2000) *Guide to Ownership, Operation and Maintenance of Building Services*. London: Chartered Institute of Building Services Engineers.

CIBSE (2005a) *AM10, Natural Ventilation in Non-domestic Buildings*. London: Chartered Institute of Building Services Engineers.

CIBSE (2005b) *CIBSE Guide A: Environmental Design*. London: Chartered Institute of Building Services Engineers.

CIBSE (2006a) *CIBSE Guide A: Environmental Design*. London: Chartered Institute of Building Services Engineers.

CIBSE (2006) *Energy Assessment & Reporting Methodology*. Second Edition. London: Chartered Institute of Building Services Engineers.

CIBSE (2006) *Renewable Energy Sources for Buildings, TM 38*. London: Chartered Institute of Building Services Engineers.

Clark, L.A. and Watson, D. (1988) Mood and the mundane: relationships between daily events and self-reported mood. *Journal of Personality and Social Psychology*, 54: 296–308.

Clements-Croome, D.J. (2000a) Computers and health in the work place. *Proceedings of Healthy Buildings Conference*, 6–10 August, University of Technology, Helsinki, 1: 119–124.

Clements-Croome, D.J. (ed.) (2000b) *Creating the Productive Workplace*. London: Spon-Routledge (Second Edition, 2006).

Clements-Croome, D.J. (2004a) *Intelligent Buildings: Design, Management and Operation*. London: Thomas Telford.

Clements-Croome, D.J. (2004b) *Electromagnetic Environments and Health in Buildings*. London: Spon Press.

Clements-Croome, D.J. (2006) *Creating the Productive Workplace*. Second Edition. Oxford: Taylor & Francis.

Clements-Croome, D.J. (2008) Work performance, productivity and indoor air, *Scandinavian Journal of Work Environment and Health Supplement*, (4):69–78.

Clements-Croome, D.J., Awbi, H.B., Bako Biro, Zs., Kochhar, N. and Williams, M. (2008) Ventilation rates in schools, *Building and Environment*, 43(3): 362–367.

Clements-Croome, D.J., Jones, K., John, G. and Loy, H. (2003) Through-life business modelling for sustainable architecture. *Proceedings of CIBSE/ASHRAE Conference*, Edinburgh, 24–26 September 2003.

Clements-Croome, D.J. and Li, B. (1995) *Impact of Indoor Environment on Productivity*, Workplace Comfort Forum, RIBA, London.

Clements-Croome, D.J. and Li, B. (2000) Productivity and indoor environment. *Proceedings of Healthy Buildings Conference*, 6–10 August 2000, University of Technology, Helsinki, Vol. 1: 629–634.

Clements-Croome, D.J., Wu, S. and John, G. (2007) *High Quality Building Services – Based on Whole Life Value*, University of Reading.

Clements-Croome *et al.* (2009) *Master Planning for Sustainable Liveable Cities*. Sixth International Conference on Green and Efficient Building and New Technologies and Products Expo. Beijing, Ministry of Construction, 29 March.

Clift, M. and Bourke, K. (1999) *Study on Whole Life Costing*. BRE report no CR 366/98, Watford: Building Research Establishment.

Construction Clients Forum (2000) *Whole-life Costing: A Client's Guide*. London.

Construction Industry Council/Strategic Forum for Construction (2005) *Selecting the Team* [Online] Available at: www.cic.org.uk/services/SelectingtheTeam.pdf.

Conti, F. (1978) *Architecture as Environment*. London: Cassell.

Cooper, C. (2001) *Conquer your Stress*. London: CIPD Books.

Courtney, R. (1992) *Environment: Keynote Address*. In: CIB World Building Congress, 18–22 May 1992, Montreal, Canada.

CRISP (1999) *UK Study on Whole-Life Costing*. Construction Research and Innovation Strategy Panel Performance Theme Group (www.crisp-uk.org.uk).

Croome, D.J. (1990) Building services engineering: the invisible architecture, *Building Services Engineering Research and Technology*, 11(1): 27–31.

Daly, S. (2010) Ecobuild Conference at Earls Court London and Personal Communication (Heath Avery).

DEFRA (2002) *Achieving a Better Quality of Life*. Government annual report on sustainable development, Department for Environment, Food and Rural Affairs.

De Marco, T. and Lister, T. (1987) *Peopleware: Productive Projects and Teams*. New York: Dorset House Publishing.

DETR (1999) *A Better Quality of Life: Strategy for Sustainable Development for the United Kingdom*. London: HMSO.

DETR (2000) *Building a Better Quality of Life: a Strategy for More Sustainable Construction*. London: HMSO.

DTI (2001) *Constructing the Future*. The Built Environment and Transport Foresight Panel, Department of Trade and Industry. London: HMSO.

DTI (2002) *The State of the Construction Industry*. Department of Trade and Industry report. London: HMSO.

DTI (2012) *Energy Consumption in the United Kingdom*, Ch. 5, Services sector energy consumption. Department of Trade and Industry, HMSO. Available at: https://www.gov.uk/government/uploads/system/uploads/attachment_data/file/65 949/file11250.pdf [accessed 28 January 2013].

Duangsuwan, J. and Liu, K. (2008) Multi-agent Control of Shared Zones in Intelligent Buildings. *Proceedings of International Conference on Computer Science and Software Engineering*, 12–14 December. Vol. 1: 1238–1241.

Earth Pledge Foundation (2005) *Green Roofs: Ecological Design and Construction*. Atglen, PA: Schiffer Publishing, Inc.

Eden Brown (2001) *Employment Survey*. Fourth Annual Eden Brown Employment Survey.

Edwards, B. (2002) *Rough Guide to Sustainability*. London: RIBA Publications.

Edwards, S., Barlett, E. and Dickie, I. (2000) *Whole Life Costing and Life Cycle Assessment for Sustainable Building Design*. Digest 452. Watford: Building Research Establishment.

Egan, J. (1998) *Rethinking Construction*. DETR, UK. [Online] Available at: http://www.constructingexcellence.org.uk/pdf/rethinking%20construction/rethin king_construction_report.pdf.

Eichholtz, P., Kok, N. and Quigley, J. (2009) *Doing Well by Doing Good? An Analysis of the Financial Performance of Green Office Buildings in the USA*. London: RICS [Online] Available at: http://www.rics.org/site/download_feed. aspx?fileID=20&fileExtension=PDF (accessed 8 October 2009).

El-haram, M., Marenjak, S. and Horner, R. (2001) Practical application of RCM to local authority housing: a pilot study, *Journal of Quality in Maintenance Engineering*, 8(2): 135–143.

El-haram, M., Marenjak, S. and Horner, R. (2001) *The Use of ILS Techniques in the Construction Industry*. MIRCE Academy Symposium, University of Exeter.

EMSD, 2003, Territory-wide implementation study for water-cooled air conditioning systems in Hong Kong. Available at: www.emsd.gov.hk/emsd/e_download/ pee/wacs_tws_es_eng.pdf [accessed December 2008].

Engen, T. (1991) *Odor Sensation and Memory*. New York: Praeger.

EPA (US Environmental Protection Agency), 2007, Emissions factor: EPA's Climate Change Action Plan (CCAP). Available at: www.epa.gov/cleanenergy/energy-resources/refs.html [accessed January 2009].

EPBD (2002) Directive (EPBD) 2002/91/EC of the European Parliament and of the Council of the 16 December 2002 on the energy performance of buildings.

Evans, J., Haryott, R., Haste, N. and Jones, A. (1998) *The Long Term Costs of Owning and Using Buildings*. London: The Royal Academy of Engineering.

Everett, R. (2009) The 'Building Colleges for the Future' program: delivering a green and intelligent building agenda, *New Review of Information Networking*, 14(1): 3–20, Routledge.

Faber Maunsell, 2002, *Tall Buildings and Sustainability Report*, Corporation of London, London, UK.

Faber Maunsell, 2004, *London Renewables*, London Energy Partnership, London, UK.

Finkelstein, W. (1988) *Integrated Logistic Support: The Design Engineering Link*. Bedford: IFS.

Fisk, W.J. (1999) Estimates of potential nationwide productivity and health benefits from better indoor environments: An update. In J.D. Spengler, J.M. Samet and J.F. McCarthy (eds) *Indoor Air Quality Handbook*. New York: McGraw-Hill, Chapter 4.

Fisk, W.J. (2000a) Health and productivity gains from better indoor environments and their relationship with building energy efficiency. *Annual Review of Energy Environment*, 25(1): 537–566.

Fisk, W.J. (2000b) Review of health and productivity gains from better IEQ. *Proceedings of Healthy Buildings*, Helsinki, 4: 24–33.

Flanagan, R. and Jewell, C. (2005) *Life Cycle Costing for Construction*. London: Surveyors Publications.

Flanagan, R. and Jewell, C. (2005) *Whole Life Appraisal for Construction*. Oxford: Blackwell Publishing.

Focus (2009) Issue 12, ISSN 1752–7473, June. Materials KTN.

Gann, D.M., Salter, A.J. and Whyte, J.K. (2003a) The design quality indicators: a tool for thinking. *Building Research & Information*, 31(5): 318–333.

Gann, D.M. and Whyte, J.K. (2003b) *Design Quality* special issue. *Building Research & Information*, 31(5).

Goleman, D. (2009) *Ecological Intelligence*. London: Allen Lane.

Graham, C.I. (2008) High Performance HVAC. *Whole Building Design Guide*. UK: National Institute of Building Sciences. [Online] Available at: http://www.wbdg.org/resources/hvac.php [accessed 26 May 2009].

Gray, C. and Flanagan, R. (1989) *The Changing Role of Specialist and Trade Contractors*. London: Chartered Institute of Building.

Greater London Authority, 2004, *The London Plan: Spatial Development Strategy for Greater London*, Greater London Authority, London, UK.

Greater London Authority, 2009, *Draft Revised Supplementary Planning Guidance: London View Management Framework: The London Plan (Spatial Development Strategy for Greater London)*, Greater London Authority, London, UK.

Green, S., Newcombe, R., Fernie, S. and Welle, S. (2004) *Learning Across Business Sectors: Knowledge Sharing Between Aerospace and Construction*. University of Reading, ICRC Report.

Gruneberg, S. (2000) *The Growth and Survival of Firms in the Heating and Ventilating Industry*. A PhD thesis submitted to the Faculty of the Built Environment, The Barlett School of Graduate Studies, University College London.

Gulliver, S. and Liu, K. (2013) Designing intelligent pervasive spaces for living and working. In D.J. Clements-Croome (ed.) *Intelligent Buildings: Design, Management and Operation*. London: Telford ICE Publishing, Chapter 8.

Haghighat, F. and Donnini, G. (1999) Impact of psycho-social factors on perception of indoor air environment studies in 12 office buildings, *Building and Environment*, 34: 479–503.

Hedge, A. (1994) Sick building syndrome: is it an environment or a psychological phenomenon? *La Riforma Medica*, 109, Supp.1 (2): 9–21.

HEEPI and SUST (2008) *High Performance Buildings. 1. The Business Case for Universities and Colleges*. Published by HEEPI-SUST-The lighthouse on sustainability.

Heerwagen, J.H. (1998) Productivity and well-being: what are the links? *American Institute of Architects Conference on Highly Effective Facilities*, 12–14 March, Cincinnati, OH.

Herzberg, F. (1966) *Work and the Nature of Man*. New York: World Publishing Company.

Heschong, L. (1979) *Thermal Delight in Architecture*. Cambridge, MA: MIT Press.

Himanen, M. (2004) The intelligence of intelligent buildings. In D. Clements-Croome (ed.) *Intelligent Buildings: Design, Management and Operation*. London: Thomas Telford.

Hirigoyen, J. (2009) *Trends in Responsible Property Investment*. Jones Lang LaSalle Inc.

Hirigoyen, J. and Newell, G. (2009) Developing a socially responsible property investment index for UK property companies. *Journal of Property Investment & Finance*, 27(5): 511–521. Emerald Group Publishing Limited.

Hodges, A. (2012) *Alan Turing: The Enigma*. London: Vintage.

House of Commons: Education and Skills Committee (2007) *Sustainable Schools: Are We Building Schools for the Future?* London: The Stationery Office. [Online] Available at: http://www.publications.parliament.uk/pa/cm200607/cmselect/cmeduski/140/140.pdf [accessed 31 July 2009].

Hughes, W., Ancell, D., Gruneberg, S. and Hirst, L. (2004) Exposing the myth of the 1:5:200 ratio relating initial cost, maintenance and staffing costs of office buildings. In F. Khosrowshahi (ed.) *20th Annual ARCOM Conference*, 1–3 September 2004, Heriot Watt University. Association of Researchers in Construction Management, Vol. 1, 373–381.

Huppert, F.A., Baylis, N. and Keverne, B. (2005) *The Science of Well-Being*. Oxford: Oxford University Press.

Hurst, R., Williams, B. and Lay, M. (2005) *Whole-Life Economics of Building Services*. International Facilities and Property Information.

IEA (2004) Annex 44, *Integrating Environmentally Responsive Elements in Buildings: A State of the Art Review*, EU.

IEA (2008) *Integrating Environmentally Responsive Elements in Buildings: Volume 1, A State of the Art Report*. International Energy Agency, Aalborg University, Denmark.

IEEE (1990) *IEEE Standard Computer Dictionary: A Compilation of IEEE Standard Computer Glossaries*. New York: Institute of Electrical and Electronics Engineers.

Innovest (2002) *Energy Management and Investor Returns: The Real Estate Sector*. Innovest Strategic Value Advisors, Inc., p. 18.

Isen, A.M. (1990) The influence of positive and negative effect on cognitive organisation: some implications for development. In N. Stein, B. Leventhal and B. Trabasso (eds) *Psychological and Biological Approaches to Emotion*. Hillsdale, NJ: Erlbaum.

ISO 1400 (1996) *International Standard for Environmental Management*. International Organisation for Standardisation.

ISO 9000 and 9001 (2000) *International Standard for Quality Management*. International Organisation for Standardisation.

JISC (2009) *Draft Strategy for 2010–2012* [Online] Available at: http://www.jisc.ac.uk/media/documents/aboutus/strategy/jisc%20strategy%20updated%20draft%202010-2012.pdf [accessed 30 July 2009].

John, G. and Clements-Croome, D.J. (2005) Innovative approach to building systems integration problems: using systems theory, technological forecasting and scenario planning. *Proceedings of the Third Innovation in Architecture, Engineering and Construction Conference* (AEC 2005), Amsterdam, Netherlands, 14–18 June, pp. 385–394.

John, G., Clements-Croome, D.J., Lo, H. and Fairey, V. (2005) Contextual prerequisites for the application of ILS principles to the building services industry, *Journal of Engineering, Construction and Architectural Management*, 12(4): 307–328.

John, G., Loy, H., Clements-Croome, D.J., Fairey, V. and Neale, K. (2003) Ageing gracefully: how can a whole life support services framework enhance the life of the building services systems? *Proceedings of the Healthy Buildings Conference*, 7–11 December 2003, National University of Singapore. Vol. 3, pp. 417–422.

Johnson, E. (2007) Building IQ: intelligent buildings are becoming part of global real estate market, *Journal of Property Management*, May. [Online] Available at: http://www.highbeam.com/doc/1G1-164222376.html [accessed 8 February 2009].

Jones, P. (1995) Health and comfort in offices. *The Architects Journal*, 8 June: 33–36.

Juniper, B.A., White, N. and Bellamy, P. (2009) Assessing employee well-being: is there another way?, *International Journal of Workplace Health Management*, 2(3): 220–230.

Kamon, E. (1978) Physiological and behavioural responses to the stresses of desert climate. In G. Golany (ed.) *Urban Planning for Arid Zones*. New York: Wiley.

Kelly, N. (2008) *Smart Buildings help NG Bailey to Cut Carbon*. Business Green. June. Available at: http://www.computing.co.uk/computing/news/2219427/smart-buildings-help-ng-bailey [accessed 8 Feb 2009].

Kuo Wand Prasad, V. (2000) An annotated overview of system reliability optimization. *IEEE Transactions on Reliability*, 49: 176–187.

Kusters, J. and Heemstra, F. (2001) Software maintenance: an approach towards control. *IEEE International Conference on Software Maintenance*, IEEE Computer Society, pp 667–670.

Latham, M. (1993) *Trust and Money*. London: HMSO.

Latham, M. (1994) *Constructing the Team: Joint Review of Procurement and Contractual Arrangements in the United Kingdom Construction Industry: Final Report*. London: HMSO.

Leaman, A. and Bordass, B. (1999) Productivity in buildings: the killer variables, *Building Research and Information*, 27(1): 4–19.

Leaman, A. and Bordass, W. (2005) Productivity in buildings: the killer variables. In D. Clements-Croome (ed.) *Creating the Productive Workplace*, Oxford: Taylor & Francis, Chapter 10.

Le Doux, J. (1996) *The Emotional Brain*. New York: Simon and Schuster.

Li, B. (1998) *Assessing the Influence of Indoor Environment on Self Assessed Productivity in Offices*. PhD Thesis, University of Reading.

Libeskind, D. (2002) The walls are alive, *Guardian*, 13 July.

Liu, K., Lin, C. and Qiao, B. (2008) A multi-agent system for intelligent pervasive spaces. *Proceedings of IEEE International Conference on Service Operations and Logistics, and Informatics (SOLI)*, pp. 1005–1010.

Liu, K., Nakata, K. and Harty, C. (2009) Pervasive informatics: theory, practice and future directions, *Journal of Intelligent Buildings International* (submitted).

London Hazards Centre (1990) *Sick Building Syndrome: Causes, Effects and Control*. London: London Hazards Centre. Available at: http://www.lhc.org.uk/members/pubs/books/sbs/sb_toc.htm [accessed 3 August 2009].

London Planning Advisory Committee, 1998, *High Buildings and Strategic Views in London*, London Planning Advisory Committee, London, UK.

Low Carbon Construction IGT (2010) *Emerging Findings* [Online] Available at: http://www.bis.gov.uk/assets/BISCore/business-sectors/docs/l/10-671-low-carbon-construction-igt-emerging-findings.pdf

Low Carbon Construction IGT (2010) *Final Report* [Online] Available at: http://www.bis.gov.uk/assets/biscore/business-sectors/docs/l/10-1266-low-carbon-construction-igt-final-report.pdf.

Loy, H., John, G., Clements-Croome, D.J., Fairey, V. and Neale, K. (2004) Achieving quality through better prediction for building services systems, *Building Services Engineering Research and Technology Journal*, 24(4): 323–333.

LSC (2009) Design Guidance Website. [Online] Available at: http://designguidance.lsc.gov.uk/NR/exeres/E8F8BCD1-F9BD-440A-B58B-DDC0AC20FDC2.htm [accessed 30 July 2009].

Mahdavi, A. (2006) The technology of sentient buildings, *ITU AlZ*, 3(1/2): 24–36.

Mao, W., Clements-Croome, D. and Mao, L. (2007) A sense diary system for intelligent buildings. *Proceedings of Clima 2007 WellBeing Indoors*, Helsinki.

Martin, A. (2003) *BMS Maintenance Guide: Plus a Model Maintenance Specification (BG 4/2003)*. London: BSRIA.

Martin, J. and McClure, C. (1983) *Software Maintenance: The Problem and its Solutions*. New Jersey: Prentice Hall.

Maslin, M. (2004) *Global Warming*. Oxford: Oxford University Press.

Maslin, M. (2007) *Global Warming*. Grantown-on-Spey, Scotland: Colin Baxter Photography Ltd.

Maslow, A.H. (1943) A theory of human motivation, *Psychology Review*, 50: 370–396.

Mawson, A. (2002) The workplace and its impact on productivity. In Series 12 *Advanced Working Papers*. London: Advanced Workplace Associates Ltd, Chapter 4.

Mayo, E. (1945) *The Social Problems of an Industrial Civilisation*. MA Thesis, Harvard University School of Business.

McAra, B. (2006) *Review the Relationship between Building Design, Cost and Quality in the Further Education Sector.* Coventry: LSC. Available at: http://readingroom.lsc.gov.uk/Lsc/2006/research/commissioned/nat-reviewthe relationshipbetweenbuildingdesigncostandqualityinthefesector-re-jun2006.pdf [accessed 17 March 2009].

McDougall, G., Kelly, J., Hinks, J. and Bititci, U. (2002) A review of the leading performance measurement tools for assessing buildings, *The Journal of Facilities Management*, 1(2): 142–153. MCB UP Ltd.

Mendell, M.J. and Smith, A.H. (1990) Consistent pattern of elevated symptoms in air-conditioned office buildings: a reanalysis of epidemiologic studies, *American Journal of Public Health*, 80(10): 1193–1199.

Microsoft Corporation (2005) *Business Case for Tablet PCs for the Pharmaceutical Industry White Paper.* Available from: http://www.microsoft.com/en-gb/download [accessed 20 November 2012].

Miller, N.G., Pogue, D., Gough, Q.D. and Davis, S.M. (2009) Green buildings and productivity, *Journal of Sustainable Real Estate*, 1(1): 65.

Ministry of Defence (1996) *Integrated Logistic Support.* Defence Standard 00–60, issue 1. Glasgow: MoD.

Ministry of Defence (1998) *Logistic Support Analysis (LSA) and Logistic Support Analysis Record (asset register).* Defence Standard 00–60 part 1, issue 2, 31 March. Glasgow: MoD.

Ministry of Defence (2002) *Application of Integrated Logistic Support (ILS).* Defence Standard part 0, issue 5, 24 May. Glasgow: MoD.

Ministry of Defence (2002) *Guide to the Application of LSA and Asset Register.* Defence Standard 00–60, part 2, issue 5, 24 May. Glasgow: MoD.

Ministry of Defence (2002) *Electronic Documentation.* Defence Standard 00–60, part 10, issue 5, 24 May. Glasgow: MoD.

Ministry of Defence (2002) *Application of Integrated Supply Support Procedures.* Defence Standard 00–60, part 20, issue 6, 24 May. Glasgow: MoD.

Mootanah, D. (2005) Researching whole life value methodologies for construction. In F. Khosrowshahi (ed.) *21st Annual ARCOM Conference*, 7–9 September 2005, SOAS, University of London. Association of Researchers in Construction Management, 2: 1247–1255.

Moubray, J. (1996) *RCM-II-Reliability-centred Maintenance.* Oxford: Butterworth-Heinemann.

Murphy, J., 2002, 'Using CO_2 for demand-controlled ventilation', *Trane Engineers Newsletter* 31(3). Available at: www.trane.com/Commercial/library/vol31_3/ [accessed 7 December 2008].

National Audit Office (2001) *Modernising Construction* [Online] Available at: http://www.nao.org.uk/publications/0001/modernising_construction.aspx.

National Audit Office (2005) *Improving Public Services through Better Construction* [Online] Available at: www.nao.org.uk/publications/0405/improving_public_services.aspx?

Nedved, M. (2011) Ventilation and the air ion effect in the indoor environments: impact on human health and well-being. In S.A. Abdul-Wahab (ed.) *Sick Building Syndrome: in Public Buildings and Workplaces* Berlin: Springer Verlag, Chapter 28.

NEMA (1989) *Lighting and Human Performance: A Review.* A Report Sponsored by the Lighting Equipment Division of the National Electrical Manufacturers Association, Washington DC, and the Lighting Research Institute, New York.

Newell, G. (2009) Developing a socially responsible property investment index for UK property companies. *Journal of Property Investment & Finance*, 27(5): 511–521.

Newman, M. (2010) Get happy, and get on with it. *Times Higher Education*, 21 January: 34–36.

Newman, P., Kenworthy, J.R., 1999, *Sustainability and Cities: Overcoming Automobile Dependence*, Island Press, Washington, DC.

Nicol, F., Humphreys, M. and Roaf, S. (2012) *Adaptive Thermal Comfort: Principles and Practice*. Oxford: Routledge.

Niemalä, R. *et al.* (2001) *Assessing the effect of the indoor environment on productivity*. Presented at the 7th REHVA World Congress, Clima 2000, Naples, 15–18 September.

Niemalä, R. *et al.* (2002) The effect of air temperature on labour productivity in call centres, *Energy and Buildings*, 34: 759–764.

Norwegian Building Research Institute (2002) Sustainable building 2002. Trine Dyrstad Pettersen (ed.) *Proceedings of the Third International Conference on Sustainable Building*, September 2002.

Noy, P., Liu, K., Clements-Croome, D.J. and Qiao, B. (2007) Design issues in personalising intelligent buildings. *Proceedings of 2nd International Conference on Intelligent Environments*, Athens, 5–6 July, IET.

Office of the Deputy Prime Minister (2006) *Part L of the Building Regulations*. Available at: http://www.safety.dtlr.gov.uk/bregs/brpub/letters/br06ab.htm.

OGC (2002) *Whole Life Costs*. Construction Procurement Guidance no. 7, Office of Government Commerce, London.

OGC (2003) *Achieving Excellence in Construction* guidance [Online] Available at: http://webarchive.nationalarchives.gov.uk/20110822131357/http://www.ogc.gov .uk/ppm_documents_construction.asp.

OGC (2005) *Common Minimum Standards* [Online] Available at: http://web archive.nationalarchives.gov.uk/20110601212617/http://www.ogc.gov.uk/docu ments/Common_Minimum_Standards_PDF.pdf. Since updated: http://www. cabinetoffice.gov.uk/sites/default/files/resources/Government_Construction_Com mon_Minimum_Standards.pdf

OGC (2007) *Guide to Best Fair Payment Practices* [Online] Available at: http://www. b-es.org/b-es-connections/commercial-and-legal/procurement/ogc-guide-to-best-fair-payment-practices/.

Oh, S.Y.J. (2005) *Indoor Air Quality and Productivity in Offices in Malaysia*. Dissertation for Bachelor's Degree, School of Construction Management and Engineering, University of Reading.

Ong, B.L. (2013) *Beyond Environmental Comfort*. Oxford: Taylor & Francis.

Pacheco-Torgal, F. and Labrincha, J.A. (2013) The future of construction materials research and the seventh UN Millenium Development Goal: a few insights, *Construction and Building Materials*, 40: 729–737.

Pacheco-Torgal, F. *et al.* (2013) *Nanotechnology in Eco-efficient Construction*. Cambridge: Woodhead Publishing.

Page-Jones, M. (1988) *The Practical Guide to Structured Systems Design*. Second edition. Englewood Cliffs, NJ: Prentice Hall.

Parsloe, C. (1997) *Allocation of Design Responsibilities for Building Engineering Services*. Technical Note TN 21/97, Building Services Research and Information Association.

Passivent Ltd (2005) *Mixed mode cooling systems*. Brochure, UK.

Pham, H. and Wang, H. (1997) Imperfect maintenance, *Quality Control and Applied Statistics*, 42: 225–226.

Poirazis, H. (2004) *Double Skin Facades for Office Buildings: A Review*, Report EBD-R-04/3, Lund Institute of Technology, Lund University, Sweden.

Pout, C.H., MacKenzie, F. and Bettle, R. (2002) *Carbon Emissions from UK Non-Domestic Buildings 2000 and Beyond*. BRE Report 442, UK.

PriceWaterhouseCoopers LLP (2008) *Building the Case for Wellness*. http://www.dwp.gov.uk/docs/hwwb-dwp-wellness-report-public.pdf [accessed 30 August 2012].

Pride, A. (2004) *Reliability Centered Maintenance*. National Institute of Building Sciences (updated 14 March 2007) http://www.wbdg.org/resources/rcm.php. Also see Alan Pride at Facilities Engineering and Operations at the Smithsonian Institute, Washington, DC.

Public Contracts Regulations (2006) No. 5 Part 1 [Online] http://www.legislation.gov.uk/uksi/2006/5/pdfs/uksi_20060005_en.pdf

Qiao, B., Liu, K. and Guy, C. (2006) A multi-agent system for building control. *Proceedings of IEEE/WIC/ACM International Conference on IAT*, December 2006, Hong Kong.

Qiao, B., Liu K., and Guy, C. (2007) Multi-agent building control in shared environment. *Proceedings of the 9th International Conference on Enterprise Information Systems*, 12–16 June 2007, Madeira, Portugal.

Quartermaine, R. (2009) *Sustainability Costs: Refurbishment, IPF Research*, Investment Property Forum, UK, January.

Ratcliff, R. (2008) *Intelligent Building Technology Can Deliver Up to 40% Energy Savings*. Intelligent Building Design. April. Available at: http://www.energy online.net/stories/articles/-/energy_management/building_controls/intelligent_ building_design/ [accessed 8 February 2009].

Rehm, M. and Ade, R. (2013) Construction costs comparison between green and conventional office buildings, *Building Research and Information*, 41(2): 198–208.

RIBA (2008) *Principles of Low Carbon Design and Refurbishment*. Royal Institute of British Architects, UK.

Satish, U. *et al.* (2011) Impact of CO_2 on human decision making and productivity. *Indoor Air Conference*, 5–10 June, Austin, TX, a574.

Shapiro, S. (2009) *Valuing Green—CBRE Makes the Financial Case for Building Green*. [Online] Available at: http://www.greenbuildinglawblog.com/2009/09/ articles/valuing-greencbre-makes-the-financial-case-for-building-green/# [accessed 8 October 2009].

Shields, B. (2003) Learning's Sound Barrier, by Nina Morgan. *Newsline*, Issue 26: 10–11. EPSRC.

Skyberg, K. *et al.* (2003) Symptoms prevalence among office employees and associations to building characteristics, *Indoor Air*: 13(3): 246–252.

Specialist Engineering Alliance (2009) *Sustainable Buildings Need Integrated Teams* [Online]. Available at: http://www.secgroup.org.uk/sea.html.

Spencer, N.C. and Winch, G.M. and Construction Industry Council (2002) *How Buildings Add Value for Clients*. London: Thomas Telford Publishers.

Steemers, K. and Manchanda, S. (2010) Energy efficient design and occupant well-being: case studies in the UK and India, *Building and Environment*, 45: 270–278.

Strategic Forum for Construction (2002) *Accelerating Change* [Online] Available at: http://www.strategicforum.org.uk/pdf/report_sept02.pdf.

Strategic Forum for Construction (2003) *Integration Toolkit* [Online] Available at: www.strategicforum.org.uk.

Strategic Forum for Construction (2006) *Construction Commitments* [Online] Available at: www.cic.org.uk/strategicforum/pdf/commitments.pdf.

Strategic Forum for Construction (2007) *Profiting from Integration* [Online] Available from: www.strategicforum.org.uk (under *Reports*).

Strelitz, Z. (ed), 2005, *Tall Buildings: A Strategic Design Guide*, RIBA Publishing & BCO, London, UK.

Strelitz, Z. (2006) *Briefing for Good Design*. At the Launch of the CIBSE Intelligent Buildings Group at the Royal Society on 19 October.

Strelitz, Z., 2006, 'Sustainable urban development: the role of tall buildings', in Department of Science and Technology, Ministry of Construction, P.R.C. Editor Committee (ed), *Memoirs On Intelligent Building and Green Building*, Vol. 2. China Architecture & Building Press, Beijing, China, 213–222.

Strelitz, Z., 2008, *Buildings That Feel Good*, RIBA Publishing, London, UK.

Su, Y., Riffat, S.B., Lin, Y. and Khan, N. (2011) Experimental and CFD study of ventilation flow rate of a Monodraught™ windcatcher, *Energy and Buildings*, 40(6): 110–116.

Tamás, G., Clements-Croome, D. and Wu, S. (2007) The use of wireless data communication and body sensing devices to evaluate occupants' comfort in buildings. *Proceedings of Clima 2007, WellBeing Indoors*, Helsinki.

Thomas, K. (2009) *Strategic Overview: Managing Environmentally Sustainable ICT in Further and Higher Education*. Bristol: JISC.

Thompson, B. and Jonas, D. (2008) Workplace design and productivity: are they inextricably interlinked? *Property in the Economy Report*, RICS: 4–41.

Tizard, G. and Mockford, J. (2008). *New Build: Delivering IT*. Available at: http://info.rsc-eastern.ac.uk/files/events/_883_DoncasterCollegeTizardMockford.ppt [accessed 18 March 2009].

Ulrich, R.S. (1984) View through a window may influence recovery from surgery, *Science*, 224: 420–421.

University of Nottingham (2011) *Post Occupancy Evaluation Report: Amenities Building, International House, Sir Colin Campbell Building, QTC Projects Ltd, Cheshire*. Available from: http://www.nottingham.ac.uk/estates/documents/poeamenitiesinternationalsircolin.pdf [accessed 31 August 2012].

USDAW (2006) *The Guardian*, Work section, 8 July 2006. Available at: http://www.guardian.co.uk/theguardian/2006/jul/08/work [accessed 30 August 2012].

US Department of Defence (1983) *Logistic Support Analysis*. Military Standard 1388–1A.

Veitch, J.A. and Galasiu, A.D. (2012) The physiological and psychological effects of windows, daylight, and view at home: review and research agenda (No. IRC-RR-325). Ottawa, ON: NRC Institute for Research in Construction.

Wang, H. (2002) A survey of maintenance policies of deteriorating systems, *European Journal of Operational Research*, 139: 469–489.

Wang, S.K. (2001) *Handbook of Air Conditioning and Refrigeration*. Second Edition. New York: McGraw Hill, Inc.

Wargocki, P. (2007) *Improving Indoor Air Quality Improves the Performance of Office Work and Schoolwork*. Technical University of Denmark, Kgs. Lyngby, Denmark [Online] Available at: http://www.inive.org/members_area/medias/pdf/Inive%5CIAQVEC2007%5CWargocki_2.pdf [accessed 10 August 2009].

Wargocki, P. and Wyon, P. (2007) The effect of moderately raised classroom temperatures and classroom ventilation rate on the performance of schoolwork by children. *HVAC&R Research*, 13(2): 193–220.

Wargocki, P., Seppanen, O., Andersson, J., Boerstra, A., Clements-Croome, D., Fitzner, K. and Hanssen, S.O. (2006) *Indoor Climate and Productivity in Offices*. Federation of European Heating and Air-conditioning Associations (REHVA) Guidebook no. 6.

Warr, P. (1998a) What is our current understanding of the relationships between well-being and work? *Journal of Occupational Psychology*, 63: 193–210.

Warr, P. (1998b) Well-being and the workplace. In D. Kahneman, E. Diener and N. Schwarz (eds) *Foundations of Hedonic Psychology, Scientific Perspectives on Enjoyment and Suffering*. New York: Russell-Sage.

Weiss, M.L. (1997) PhD Thesis, Division of Behaviours and Cognitive Sciences, Rochester University, New York.

WFC, 2009, The Intelligent Building Index (IBI), Assessment on World Financial Centre Beijing, November, Hong Kong.

Williams, B. (2006) Building performance: the value management approach. In D.J. Clements-Croome (ed.) *Creating the Productive Workplace*, Oxford: Taylor & Francis, Chapter 27.

Williams, B. (2007) MSc Presentation Course Notes – October 2007 Building Quality Assessment Method.

Wolstenholme, A. and Constructing Excellence (2009) *Never waste a good crisis* [Online] Available at http://www.constructingexcellence.org.uk/pdf/Wolstenholme_Report_Oct_2009.pdf.

Woods, J. (1989) Cost avoidance and productivity in owning and operating buildings. In J. Cone and M. Hodgson (eds) *Occupational Medicine: State of the Art Reviews*, 4 (4); and also *Problem Buildings: Buildings Associated Illness and the Sick Building Syndrome*, Philadelphia; Hanley and Belfus, 753–770.

Wu, S. and Clements-Croome, D.J. (2005) Burn-in policies for products having dormant states. *Proceedings of the Fourth International Conference on Quality and Reliability, Beijing*, pp. 567–574.

Wu, S. and Clements-Croome, D.J. (2005) Preventive maintenance models with random maintenance quality, *Reliability Engineering & System Safety*, 90(1): 99–105.

Wu, S. and Clements-Croome, D.J. (2005a) Critical reliability issues for building services systems. *Proceedings of the Fourth International Conference on Quality and Reliability*, Beijing, pp. 559–66.

Wu, S. and Clements-Croome, D.J. (2005b) Optimal maintenance policies under different operational schedules, *IEEE Transactions on Reliability*, 54(2): 338–346.

Wu, S. and Clements-Croome, D.J. (2006) A novel repair model for imperfect maintenance, *IMA Journal of Management Mathematics*, 17(3): 235–243.

Wu, S., Clements-Croome, D.J., Fairley, V., Albany, B., Sidhu, J., Desmond, D. and Neale, K. (2006) Reliability in the whole life cycle of building systems, *Engineering Construction and Architectural Management*, 13: 136–153.

Wu, S. and Clements-Croome, D.J. (2007) Burn-in policies for products having dormant states, *Reliability Engineering and System Safety*, 92(3): 278–285.

Wyon, D. (1996) Indoor environmental effects on productivity. *Keynote address in Indoor Air 1996, Paths to Better Building Environments*, Atlanta, GA: ASHRAE, 5–15.

Yeang, K., 2002, *Reinventing The Skyscraper: A Vertical Theory of Urban Design*, Wiley-Academy, Chichester, UK.

ZZA, 2002, *Tall Office Buildings in London: Giving Occupiers a Voice, British Council for Offices*, London, UK.

ZZA, 2009, *North Kent Police Station, Post Occupancy Evaluation*, Design Policy, Home Office, London, UK.

ZZA, 2010, *Bodmin Strategic Operational Policing Hub, Post Occupancy Evaluation*, Design Policy, Home Office, London, UK.

Websites

http://www.breeam.org/page_1col.jsp?id=54
http://www.cibse.org/
http://www.nationwidefilters.com/NFC/index.htm
http://www.apollolighting.co.uk/products/Technical/Activiva%20Lamps/index.php

Index